T0258751

Volume 19

URBAN FRANCE

URBAN FRANCE

IAN SCARGILL

Routledge
Taylor & Francis Group

LONDON AND NEW YORK

First published in 1983 by Croom Helm Ltd

This edition first published in 2018
by Routledge
2 Park Square, Milton Park, Abingdon, Oxon OX14 4RN

and by Routledge
711 Third Avenue, New York, NY 10017

Routledge is an imprint of the Taylor & Francis Group, an informa business

© 1983 D.I. Scargill

All rights reserved. No part of this book may be reprinted or reproduced or utilised in any form or by any electronic, mechanical, or other means, now known or hereafter invented, including photocopying and recording, or in any information storage or retrieval system, without permission in writing from the publishers.

Trademark notice: Product or corporate names may be trademarks or registered trademarks, and are used only for identification and explanation without intent to infringe.

British Library Cataloguing in Publication Data
A catalogue record for this book is available from the British Library

ISBN: 978-1-138-49611-8 (Set)
ISBN: 978-1-351-02214-9 (Set) (ebk)
ISBN: 978-1-138-48402-3 (Volume 19) (hbk)
ISBN: 978-1-138-48413-9 (Volume 19) (pbk)
ISBN: 978-1-351-05302-0 (Volume 19) (ebk)

Publisher's Note
The publisher has gone to great lengths to ensure the quality of this reprint but points out that some imperfections in the original copies may be apparent.

Disclaimer
The publisher has made every effort to trace copyright holders and would welcome correspondence from those they have been unable to trace.

Urban France

IAN SCARGILL

CROOM HELM
London & Canberra
ST. MARTIN'S PRESS
New York

© 1983 D.I. Scargill
Croom Helm Ltd, Provident House, Burrell Row,
Beckenham, Kent BR3 1AT
Croom Helm Australia, PO Box 391, Manuka,
ACT 2603, Australia

British Library Cataloguing in Publication Data

Scargill, Ian
   Urban France.
   1. Cities and towns--France
   I. Title
   307.7'6'0944   HT135
   ISBN 0-7099-2350-3

All rights reserved. For information, write:
St. Martin's Press, Inc., 175 Fifth Avenue, New York, NY 10010
First Published in the United States of America in 1983

Library of Congress Cataloging in Publication Data

Scargill, David Ian.
   Urban France.

   Bibliography: p.
   Includes index.
   1. Urbanization--France. 2. City planning--France.
3. Cities and towns--France--Case Studies. 4. New
Towns--France--Case studies.
I. Title.
HT135.S3   1984   307.7'6'0944   83-16093
ISBN 0-312-83449-7

Printed and bound in Great Britain

CONTENTS

TABLES

FIGURES

To E.T.W.ROBINSON

Chapter 1.

POSTWAR URBANIZATION

Writing in Le Monde, in July, 1981, two newly-elected
députés suggested that the recent socialist victory in the
general election had been a consequence of the problems of
the cities.   'Le monde urbain a voté à gauche ... La ville
est malade'.   They went on to contrast the city as it used
to be, a market and a meeting-place, with the ill-planned
metropolis of the present day, its population swollen with
rural overseas migrants, a focus of mounting unemployment
and of weakened social bonds.   What could once have been
described as 'un milieu vivant' had become 'un entassement
difforme'.   Urban life was now the life of the grand ensemble
and the distant suburb, summed up by the popular phrase,
'métro - boulot - dodo', which translates roughly as, travel -
work - sleep.   The authors went on to state that the answers
to these problems were to be found only in far-reaching
economic and social reforms and in the policies of decentralization proposed by the new government.
     Political arguments aside, there are few observers who
would dispute the point made by the députés that French cities
have been experiencing rapid and, at times it has seemed,
almost uncontrolled growth over the last 30 years and that
serious consequences have followed.   The nature of this urban
explosion, and the efforts of planners and others to find
solutions to the resultant problems, together form the subject
of the present work.

The Rapidity of Urbanization
     In 1850 the proportion of the French population
officially classed as 'urban' was only 25% of the total and
it did not reach 50% until 1928, more than half-a-century
later than in Britain or the Netherlands.   In 1946, when
the urban population totalled 21.25 millions, the proportion
was still only 53.2%, but thereafter the urban component
increased rapidly in response to changing social and economic
circumstances.

Measurement of urbanization depends upon how one chooses to define the 'urban' element in a population. Until 1962 the commonly accepted definition in France was based on the proportion of the population living in communes having a total of at least 2,000 in the chef-lieu. Defined in this way, 61.6% of the population was urban in 1962 (Table 1). It is a method of calculation which tends to underestimate the total, however, excluding as it often does the fringes of a town which have spread into a neighbouring 'rural' commune. The concept of the unité urbaine, introduced in 1962, takes account of this problem by including in the definition of 'urban' the population of an agglomération even if the latter should extend over parts of more than one commune. The idea of agglomération is, in turn, based on association of buildings, none being included that are more than 200 metres from the more closely built-up areas. Using the unité urbaine, the urban proportion of the population was 63.4% in 1962; by 1975 this had risen to 72.9%.

TABLE 1. URBANIZATION

| Definition | Percentage of the population 'urban' | | | |
	1954	1962	1968	1975
Pre-1962	56.0	61.6	–	–
Unités urbaines	58.6	63.4	71.3	72.9
ZPIU	–	–	79.0	82.5

A third approach to definition is one that takes account of the occupational structure of the population, recognizing that many 'rural' communes include a non-farm element of commuters and others who work in industry or tourism. From this arises the concept of the zone de peuplement industriel et urbain (ZPIU) which embraces those localities that are predominantly 'urban' in economic activity. Using these broader areas as a basis for calculation, the 'urban' component of the French population rises to 82.5% in 1975.

It is clear from the above that there is no single satisfactory means by which to calculate the urban share of a country's population. It is hardly surprising, therefore, that different measures of urbanization are presented by different authors. Noin (1976), for example, gives a figure of 77.1% for the urban component in 1975, whilst Vaughan, Kolinsky and Sheriff (1980) suggest a proportion of 'nearly 75%'. The variations arise quite obviously from the method of calculation preferred and there is little to be gained from pursuing them. Of greater importance from the point of view of the present work is the overall picture of a country experiencing rapid and large-scale urbanization. From a

figure of little more than 50% in the late 1940s, the share of
France's population that was urban rose to around three-
quarters of the total within the next 30 years.   In absolute
terms, the population of French towns and cities was swelled
over that period by the addition of at least 16 million persons,
giving rise to a host of social problems but creating, at the
same time, a market for the rejuvenated economy.

Population Growth
      The expansion of French cities since World War II has
come about both as a result of a greatly accelerated rate of
population growth and of large-scale migration from the rural
areas and abroad.
      Pre-war France was remarkable for the country's demo-
graphic stagnation.   The roots of this situation are to be
found in the nineteenth century, especially in the last quarter
when the birth rate fell from around 26 live births per thousand population to 22, and then to 20 by 1913.   With a death
rate of around $18/19‰$ in the early years of the present century,
there was little absolute growth of population.   The Great
War resulted in the death of 1.3 million Frenchmen, and if
allowance is made for other deaths and for the births that did
not take place as a result of the war, the demographic loss to
France was of the order of 3 millions.   A brief revival after
the war, when growth was assisted by the influx of a sizeable
foreign population, was followed by a renewed fall in the
birth rate which reached a trough of $14.6‰$ in 1938.   The
number of deaths exceeded that of births every year in the late
'thirties.   Many families were childless or content to have
only one child, and in 1938 the fecundity rate for all women
aged 15 to 49 had fallen to the extraordinarily low level of
$61‰$ The death rate at that time was still $15.4‰$, a consequence of an ageing population and of problems such as alco-
holism and poor housing.   In fact mortality rates remained
high, especially amongst the poor who suffered most from the
neglect of housing and the social services in general.
      At the outbreak of war in 1939 the population of France
numbered 41.9 millions, little more than the total recorded at
the time of the census in 1911 (41,479,000, including Alsace-
Lorraine).   Losses in the Second War were appreciably less
than in 1914-18, but they nevertheless included the deaths of
some 350,000 civilians due to bombing, deportation and other
causes.   The census of 1946 recorded a population total of
only 40.5 millions.
      Considered against this demographic background, the
growth of the French population since the war has been
remarkable.   The ratio of births to married women of child-
bearing years began to rise during the war years and the birth
rate reached $21‰$ in the 'baby-boom' period of the late 1940s.
Thereafter there was a slight reduction, but the birth rate

3

was still around 18‰ at the time of the 1962 census.   The
death rate had also fallen sharply in response to improved
medical and social care and by 1962 was as low as 11‰.   As a
result of these trends, five million more births than deaths
were recorded in France between 1946 and 1962.   After many
years of what Beaujeu-Garnier (1975) has called an 'ambiance
anti-familiale', it had once more become popular to have children.   The French had not managed to respond fully to General
de Gaulle's call for 12 million babies in ten years, but the
demographic state of France had nevertheless been transformed.

Although one must be wary of attributing the change to
any single cause, there can be little doubt that it reflects
in some measure the government's <u>politique familiale</u> or
natalist policy (Dyer, 1978).   The level of family allowances
had been improved as a result of the Code de la Famille of
1939, and subsequent legislation introduced generous tax
concessions for couples with children.   By a happy chance the
parents able to take advantage of these measures were ones
born in the earlier 'baby-boom' after the Great War.   There
were more marriages, therefore, and these were producing many
more children.

The birth rate fell, but only very slowly, during the
1960s.   It had dropped below 17‰ by the end of the decade but
subsequently rose to 17.2‰ in 1971 when 878,600 births took
place, the highest number since the war.   Children born in
the early postwar years were themselves marrying and having
children and large numbers of births were recorded also in
1972 and 1973.   Yet fecundity was already declining.   There
were fewer marriages after 1972; more women went out to work;
contraceptives were now available on prescription, and legislation to permit abortion before the end of the tenth week of
pregnancy was introduced at the beginning of 1975.   The birth
rate slumped to 15.2‰ in 1974 and to 14‰ in 1975.   The death
rate had levelled out at a rate of 10.5‰ and dramatically,
within the space of a couple of years, the demographic
situation had reassumed many of the characteristics of the
interwar years.

Fewer births were recorded in France in 1976 (720,000)
than in any postwar year.   Since that time the birth rate has
risen again, notably in 1979 and again in 1980 when almost
800,000 children were born, equal to the total of 1974.
Amongst the reasons for this change is the further fall that
has taken place in the rate of infant mortality, from a
figure of around 18 per thousand births in 1970 to under 10‰
in 1980.   Yet despite the increase, the number of births in
1979-80 was still below the level necessary to ensure repro-
duction of the population and the indications were that the
birth rate would fall again in the early 1980s.   Pointers
could be found in the fact that only 334,000 marriages took
place in 1980 compared with 407,000 in 1972 (the highest post-

war figure) and that more than one in five of these marriages seemed destined to end in divorce compared with one in ten in the early 1960s.

Our concern is not, however, with the prediction of population trends in the 1980s. Even if these are towards stabilization of the total and a renewed pattern of ageing, the effects of the earlier period of postwar growth remain everywhere apparent in the towns and cities of France. The national census of March, 1982 recorded a total French population of 54.2 millions, nearly 14 million more than in 1946. It is worth recalling that in 1946 the total had been only 12.25 million more than in 1801, a century and a half earlier.

Migration

Natural increase has played by far the greater part in postwar population growth in France, but migration from outside the country accounted for between a quarter and a third of the increase between 1946 and 1975. The effect of migration was greater if one allows for the generally higher fecundity of the migrant population.

TABLE 2. INTERCENSAL POPULATION CHANGE

	Annual rate of population change		
	1954-62	1962-68	1968-75
Total	+1.1	+1.2	+0.8
By natural increase	+0.7	+0.7	+0.6
By migration	+0.4	+0.5	+0.2
of which (i) repatriates	+0.1	+0.3	+0.0
(ii) others	+0.3	+0.2	+0.2

The presence of a large foreign-born element has been a characteristic of the French population for many years. The slow rate of increase of the native population meant that jobs were available for immigrants who were attracted to work mainly in agriculture, mining and the construction industry or in domestic service. Some came seasonally but others settled in the country, encouraged by liberal attitudes towards the granting of French nationality. The census of 1911 recorded 1,133,000 foreigners in France; by 1931 the total had risen to 2,891,000, some 7% of the country's population. Italians accounted for more than a third of the foreign-born in 1911 and Belgians for a further 25%. Most had settled in the départements closest to their country of origin (Dyer, 1978). A feature of the interwar years was the influx of a large number of Poles who formed colonies in the towns and pit villages of the industrial Nord.

An average of 50,000 foreign workers a year settled in

France during the first fifteen years after the Second War
(Beaujeu-Garnier, 1975) but this figure rose steeply in the
1960s when the annual rate of arrival was closer to 150,000.
A record 174,000 foreign workers entered the country in 1970.
Not all of these settled permanently, but there was a large net
gain, and the census of 1968 counted 2,621,000 foreigners.
By 1974, when restrictions were imposed on immigration, the
total foreign-born population had risen to around 4 millions.

A marked shift took place over the course of this phase
of maximum immigration in the relative importance of the
countries sending migrants to France. The number of Italians
living in France remained fairly constant but their share of
the total fell as Italian migration virtually came to an end
during the course of the 1950s. In contrast, the contribution of Spain and Portugal rose steeply, accounting for
nearly half the arrivals in the early 1960s. Spanish movement slowed considerably after 1967 but immigration of
Portuguese continued to increase in the late 1960s and early
1970s. The main feature of this later period, however, was
the influx of large numbers from North Africa, principally
from Algeria but including also Moroccans and Tunisians.
They were joined by smaller groups from Turkey and Yugoslavia.

TABLE 3. FOREIGN-BORN POPULATION

Origin of foreign-born population, 1.1.81 (largest 10 groups)

	Total	Percentage
Portugal	857,324	20.8
Algeria	808,176	19.6
Italy	469,189	11.4
Spain	424,692	10.3
Morocco	421,265	10.2
Tunisia	181,618	4.4
Turkey	103,946	2.5
Yugoslavia	68,239	1.6
Poland	65,594	1.6
Belgium	59,968	1.5

The number of foreign-born living in France at the
beginning of 1981 was 4,124,317, 8% of the population,
according to the official statistics of the Ministry of the
Interior. This figure does not take account of clandestine
immigrants ('clandos'), the estimated total of which has been
put as high as 300,000. People of North African origin
accounted for over a third of the total foreign-born at
1 January, 1981; Portuguese for one-fifth (Table 3).

There are many more male than female migrants living in
France, but amongst those who are married, fecundity is high,
immigrants being responsible for 11-12% of the births recorded
in France in the late 1970s.   Of those aged 16 and over in
January, 1981, 2.08 million were male (64.3%) and 1.15 million
female (35.7%).   The foreign-born total also included nearly
900,000 children under the age of 16.   Foreigners now constitute a major component of the workforce in the mining industry,
in construction (40% of the total) and in unskilled occupations
They reside in largest numbers in the major cities and the
industrial towns of the north-east and Lorraine.   Of the 1981
total, 1.35 millions were living in the Paris Region, 545,000
in Rhône-Alpes (including Lyon) and 378,000 in Provence-Côte
d'Azur (including Marseille).   Figures for the Nord Region
were 218,000 and for Lorraine, 192,000.   The regions of the
north-west are those with fewest foreigners - 21,000 in
Bretagne, for example.
     Within the cities the migrants display a high degree of
clustering in certain quarters.   This is especially true of
the population of North African origin.   Bidonvilles - shanty
towns - sprang up on the outskirts of Paris and other cities
in the 1960s, but most of these have now been cleared, and the
migrants tend to live either in high-rise HLM housing
(Chapter 5) or in the older properties of certain inner city
quartiers.   In Marseille, where more than one in ten of the
population is of foreign origin (four-fifths from North
Africa), immigrants are heavily concentrated in the grands
ensembles on the northern side of the city.   In these
'concrete ghettoes' the percentage of immigrants to native-
born may be as high as 70%.   Many immigrants from North Africa
also pass through Marseille, staying for varying periods of
time, and the quartier of the Porte d'Aix has become the principal reception area for such people of passage.   In addition,
many unmarried migrants live in poor quality hostel accommodation (hôtels meublés).   Immigrants are heavily clustered
in the working-class districts of Paris such as Nanterre,
Colombes and Gennevilliers.   A quarter of Nanterre's population, for example, is of overseas origin and the children
of immigrant families account for up to three-quarters of the
total in some schools (Le Monde, 27.2.81).
     Foreign workers and their families do not account for the
whole of the increase of population by migration.   Allowance
must also be made for the many repatriates coming home to
France: French families from the former colonies in Indo-China,
Tropical Africa and, above all, North Africa.   Nearly a
million pieds noirs returned from Algeria alone after the
granting of independence in 1962.   Some of these subsequently
moved on to North America and elsewhere, but a substantial
majority settled in France, especially in the south, adding
significantly to the overall gains due to migration in the

1960s (Table 2).

Some of the pieds noirs have bought farms and settled in rural France; some of the overseas workers are also employed in agricultural occupations. No measure is available of the division of migrants between rural and urban communes. It may be safely assumed, however, that a high proportion of the migrants have settled in towns and cities and that migration from overseas has contributed an important element to the population growth of urban areas.

The above discussion has been concerned with migration from abroad. An explanation of urbanization must also take account of rural to urban migration that has taken place within France.

A third of the French working population was still employed on the land after the Second World War. By 1981 only 8% of the population lived or worked on farm holdings. The rate of loss of the agricultural population has been an accelerating one: 1.9% a year between 1955 and 1963; 2.3% 1963-71; 3.5% 1971-80. Expressed in absolute terms, the twenty years before 1975 saw 3 million people leave the land for the towns, an average annual loss of 150,000. Some of these rural migrants moved directly to Paris and the largest provincial cities. To them, urban life would not necessarily be unfamiliar, France having a long tradition of temporary migration involving the seasonal movement of masons, chimney-sweepers or pedlars from the poor regions such as the Massif Central. From some districts like the Morvan it was more common for the women to travel, in this case to serve as wet-nurses in wealthy urban households. But to other postwar migrants the town was a much less familiar environment and the initial move took many of them no further than the local market town or, at most, the departmental capital. Later moves, to the larger cities, often followed however. Such stepped migration has had important consequences for the evolution of the national urban system, as we shall see in the chapters that follow.

Chapter 2.

THE URBAN SYSTEM

Evolution of the System

Compared with countries such as Britain, which experienced the full force of the Industrial Revolution, the history of the French urban system has been until recently one of stability and of slowness to change. Once its framework had been completed in the Middle Ages, there were few new additions to the system, whilst the rank-order of urban centres making up the system exhibited little response to political and social circumstances at home. The shape of the urban network was one that reflected a 'process of urbanization working through a synthesis of traditional forces' (Bédarida, 1980).

The framework of the French urban system is largely the product of two periods of town foundation, one Gallo-Roman, the other medieval. To a sprinkling of Greek ports and native oppida, the Romans added a network of military and trading settlements. Careful choice of site ensured that most of these survived the period of urban depopulation which followed the later Germanic invasions. Indeed, several authors have commented on the extent of the continuity between Gallo-Roman and modern urban settlements, 26 out of the 44 capitals of the 'civitates' - tribal territories used for administrative purposes - later becoming chefs-lieux of départements (Pinchemel, 1969; Clout, 1977 a; Dalmasso, N.D.).

A second phase of town formation corresponds with the period of demographic and economic progress that took place between the eleventh and fourteenth centuries. The growth of trade spawned many new market towns, founded for profit by local landlords and peopled by a new generation of citizens seeking the freedom and privileges set out in their charters. Most grew around a pre-existing village nucleus, but some were founded de novo, the best known of these medieval new towns being the bastides of the south-west (Beresford, 1967). Castles, as at Saumur, or monasteries, such as Cluny, also provided catalysts to urban development. Above this basic layer of small exchange centres were the larger, and often older, towns which attracted internationally-famous fairs because of their position on important routeways. Outstanding amongst these were Lyon, and the towns of Champagne (Reims, Troyes, Châlons-sur-Marne) from which Paris drew some of its growth impetus in the late Middle Ages. Trade was linked with manufacturing in Flanders where both new and old towns benefited from an expanding textile industry.

By the middle of the fourteenth century the outline of the French urban system was substantially complete, subsequent modifications to it being brought about more as a result of the varied rates of growth or decline of the urban centres making up the system than by the addition of new ones. Exceptions to this generalization were few, Clout (1977 b) for example, stating that no more than a score of new towns were built in France between 1500 and 1800. Those towns that were founded de novo mainly owed their origin to political ambition and the need to defend frontiers. Most of them were built close to the eastern boundary. They included Rocroi, Hesdin and Villefranche-sur-Meuse in the sixteenth century and those built by Vauban between 1679 (Longwy) and 1698 (Neuf Brisach). Colonial aims led to the creation of new ports at Le Havre (1517-43) and, within a few years of each other in the 1650s and 1660s: Rochefort, Brest, Lorient and Sète. The remaining foundations were built for self-aggrandisement by king or nobles, their plans incorporating the increasingly exaggerated ideas of the Renaissance. First expressed in the rebuilding of Vitry-le-François for Francis I (1545), the search for the ideal city is to be seen in the Duc de Sully's Henrichemont north of Bourges and the Duc de Nevers's Charleville (both 1608), in the town of Richelieu, founded by the Cardinal in 1635, and in Louis XIV's Versailles (1671).

Nineteenth-century industrialization had a far less disruptive effect on the French urban system than on that of Britain. Expansion of the coal, iron and textile industries was accompanied by rapid growth of towns on or close to the northern coalfield and isolated centres elsewhere, but there was nothing to parallel the history of a Middlesbrough or a Barrow-in-Furness. The nearest equivalent was the town of Saint-Etienne which saw its population grow from 16,000 to some 200,000 during the course of the century. The effect of the railway was felt after 1850 but this served mainly to confirm the position of the larger towns in the urban hierarchy. La Roche-Migenne, on the main line from Paris to Dijon, is the French equivalent of a Wolverton or Swindon, otherwise the additions to the urban system most directly attributable to the railway were a number of spas (Vichy, Aix-les-Bains) and fashionable holiday resorts (Deauville and, later, Le Touquet). As in Britain, towns by-passed by the railway experienced relative decline; such was the case at Alençon, for example.

Paris had a population of almost 550,000 at the beginning

of the nineteenth century.   Three other cities - Marseille,
Lyon and Bordeaux - recorded totals close to 100,000 when the
first national census was held in 1801.   The next cities, in
descending order of size were: Rouen (87,000), Nantes (74,000),
Lille (55,000), Toulouse (50,000) and Strasbourg (49,000).
It is indicative of the stability that characterized the French
urban system during the nineteenth century that seven of these
eight provincial centres (Rouen was the exception) were the
places chosen by the government in 1964 to act as métropoles
d'équilibre (Chapter 3).   Only Nancy, of the métropoles was
not amongst the ten largest provincial oities in 1801.

The Revolution had substituted a system of 83 départements for the framework of provincial rule that had prevailed
under the ancien régime.   The new pattern of administrative
units represented by no means, however, a total break with the
past, the officers of the National (or Constituent) Assembly,
who carried out the division of the country, showing a considerable degree of sensitivity to existing territorial
loyalties, especially to the boundaries of dioceses.   There
are numerous parallels with what happened in Ariege, for
example, where the new département corresponded closely with
the old Comté de Foix and the diocese of Pamiers.   In some
cases the selection of a particular town as site of the
préfecture confirmed its progress at the expense of rival
centres.   This was so in Dordogne where it was originally
agreed that Périgueux, capital of historic Périgord, should
share the administration of the new département with the towns
of Sarlat and Bergerac which were of almost equal size, each
acting as chef-lieu for a period of years.   It was an
impractical solution.   Périgueux was chosen to serve as
chef-lieu first and, in fact, never gave up the role, slowly
outgrowing the other towns because of the advantages which its
administrative responsibilities conferred (Scargill, 1974).
One must not exaggerate the importance of this latter function,
however.   Division of local government between 83 préfectures
was sufficient to ensure that most departmental capitals
remained modest in size - the villes moyennes of the 1970s
(Chapter 4).   The only new towns created for political purposes were Pontivy and La-Roche-sur-Yon.

Throughout the nineteenth century the most common form of
urban settlement remained the small market town of some
5,000-20,000 inhabitants and the minor administrative and
regional centre of 20,000-50,000 (Sutcliffe, 1980).   The
impact of industrialization on the urban system was localized
but had the effect, overall, of increasing the disparities
between an expanding north and east and a traditional and still
largely rural, west and south.   In general it was the larger
provincial cities that experienced most growth, notably in the
second half of the century after the main line rail network
had been completed, but even that seemed little in comparison

with the massive concentration of population on Paris.   The
capital's expansion cast a shadow over the whole country, its
effect being most severe in the Paris Basin where it acted
increasingly as a brake on the growth of such towns as Orléans,
Rouen and Troyes.   Orléans, for example, which ranked sixteenth in size amongst French towns in 1851, had fallen to
thirtieth a hundred years later.

The Primacy of Paris

Notwithstanding the centralized nature of the French state
the growth of Paris was relatively modest up to the nineteenth
century.   Only 100,000 inhabitants were added to the capital's
population in the 150 years before the Revolution, a slower
rate of growth than that of the country as a whole.   The
court was installed at Versailles and the king rarely visited
Paris.   Indeed, fear of the mob brought deliberate attempts
to restrict the expansion of the city.   The same policy was
pursued after the Revolution and the inclusion of Paris in a
small <u>département</u> of only 480 sq.km. has been interpreted as a
conscious attempt on the part of the <u>constituants</u> to limit the
influence of the capital (Gravier, 1947).   There were more
than half-a-million people living in Paris at the beginning of
the nineteenth century but this was only 2% of the national
total.

Napoleon's accession to power reversed these earlier
attempts to control the growth of Paris.   Bonaparte wanted a
capital to match his imperial ambition and a programme of
public works was begun in order to gain international prestige
for the city.   At the same time, the establishment of <u>préfets</u>
in the <u>départements</u> consolidated the decision-making role of
central government.   Population grew, and by mid-century Paris
had over a million inhabitants.   If allowance is made for
suburban development in adjacent communes, this total may be
raised to 1,250,000, but it was still only 3-4% of the national
total.   Napoleon's policies had initiated the rise of Paris
to a position of primacy in the urban system, but the massive
growth which confirmed that position had to await the completion
of the main line rail network.

Jean-Baptiste Legrand, a director of the <u>Travaux Publics</u>,
is said to have conceived the idea of a radiating rail network,
linking the chief provincial cities with Paris, in 1832.   His
plan was approved ten years later, and by the mid-1850s this
primary network was complete.   Inspired by strategic considerations, its economic impact was enormous.   Industrial
centralization now followed political centralization, with
rapid expansion taking place, particularly in the metal fabri-
cating and chemical industries.   Migrants poured into the
capital from the provinces, the lure of employment increased
by the problems experienced in rural France: phylloxera,
competition from foreign grain, and the collapse of the

12

traditional craft industries.

Louis-Napoleon appointed Baron Haussemann préfet of the département of Seine in 1853 and the latter quickly embarked on his plan to open up the city with boulevards and places. The construction works provided more employment for the tide of migrants. They also contributed to the physical expansion of the metropolis, the poor who were dispossessed moving out to the industrial suburbs springing up beyond the main line railway stations and, later, beyond the limits of the Cité as defined by the walls of 1841-45. Industry followed the railways and the waterways to Saint-Denis, Saint-Ouen, Ivry and other places that make up the inner suburban ring of the modern city. Expansion of industrial employment was particularly marked from around 1890 with the emergence of large-scale enterprises. This continued through the First World War and thereafter when renewed impetus for growth came from the success of the vehicle industry, major car, and later aircraft, plants being set up on the flat land afforded by the gravel terraces of the Seine. By 1936, Parisian manufacturing gave employment to a fifth of the national industrial workforce. The capital was responsible for around half the jobs in vehicle and electrical construction, nearly three-quarters of those in precision engineering.

The industrial dominance of Paris was more than matched by the city's position with regard to higher order services. The influential grandes écoles were almost all in the capital, whilst the university of Paris alone accounted for nearly half the university student population of France as a whole. Banking and financial services, the press and publishing, theatres and department stores had become similarly highly centralized.

The population of Paris increased three-fold during the second half of the nineteenth century in response to the ever greater control which the capital exercised over French affairs. Overcrowded housing and unhealthy living conditions contributed to a high death rate, especially amongst the very young, so that growth was sustained largely by immigration from the provinces. Clout (1977 b) has put the contribution of immigration to population increase in the nineteenth century as high as 90% and the city developed distinctive enclaves, often close to the main line railway stations, where immigrants from different regions of France tended to cluster on arrival (Ogden and Winchester, 1975).

Until the final quarter of the nineteenth century, the physical expansion of Paris was largely contained within the line of the 1841-45 walls. Thereafter it affected increasingly the suburbs beyond, although population continued to squeeze into whatever space was available in the ville, which did not reach its maximum population of 2.9 millions until 1921. The flow of migrants continued during the early decades of the

present century so that Greater Paris had a population in
excess of 6 millions at the outbreak of the Second World War,
some 15% of the national total.   If the boundaries of the
agglomeration are defined generously, it is possible to say
that the whole of the population increase which took place in
France in the ninety years after 1850 had been absorbed by
Paris.   The primacy of Paris was established, its impact on
the rest of the country summed up in Gravier's (1947) now-
famous phrase, 'Paris et le désert français'.

Postwar Movement of Population
Before any attempt is made to define the French urban
system it is necessary to look first at the changes that have
taken place in the distribution of the French population since
the Second World War, taking account both of inter-regional
movements and of the apparent attraction to the population of
different size-categories of city.   Population changes that
have taken place in the 21 régions de programme of mainland
France during the inter-censal periods, 1954-62, 1962-68 and
1968-75, are set out in Table 4, which also shows the relative
importance of natural increase and of migration.
The contribution of natural increase to population growth
has been greater in the northern half of the country than in
the south.   This has been true throughout the period under
consideration, although an exception may be found in the north-
west where prolonged out-migration has contributed to a
lowering in the more rural areas of the once-high rate of
natural increase.   In those regions such as Nord, Lorraine
and Normandy where postwar birth rates have been particularly
high, the pressure on schools, housing and related services
has been severe.
The pattern of migratory movement is more complex than
that of natural increase, and has changed since the mid-1950s.
Then the flow was principally from rural to urban areas in
response to change on the land and the pull of new opportunities in the cities.   Rural to urban migration has not
ceased, but inter-urban movement is now more common, with
workers leaving areas of declining industry for those cities
that offer better prospects of employment.   There is also a
counter-movement from the larger cities to nearby 'rural'
communes.   Significantly the 'rural' communes with a population in excess of 1,000 experienced an overall gain by
migration between 1968 and 1975, whilst the more thinly populated ones continued to lose migrants.   Migration also
involves the movement of retired persons, an important con-
tributory factor in the growth of Nice and other towns of the
'sunbelt' of Provence - Côte d'Azur.   The average distance
over which migration takes place has tended to increase, but
there are differences between western and eastern France in
this respect, longer moves being more typical in the west

14

TABLE 4. ANNUAL RATES OF POPULATION CHANGE

	1954-1962			1962-1968			1968-1975		
	Total	Natural Increase	Migration	Total	Natural Increase	Migration	Total	Natural Increase	Migration
Alsace	+1.01	+0.74	+0.27	+1.18	+0.73	+0.46	+1.06	+0.55	+0.51
Aquitaine	+0.63	+0.37	+0.26	+1.03	+0.34	+0.69	+0.54	+0.24	+0.30
Auvergne	+0.25	+0.27	-0.02	+0.49	+0.23	+0.27	+0.21	+0.20	+0.01
Bourgogne	+0.58	+0.43	+0.15	+0.73	+0.39	+0.33	+0.64	+0.35	+0.30
Bretagne	+0.30	+0.67	-0.38	+0.48	+0.57	-0.08	+0.73	+0.51	+0.22
Centre	+0.73	+0.58	+0.15	+1.16	+0.53	+0.63	+1.12	+0.49	+0.63
Champagne-Ardennes	+0.84	+0.95	-0.11	+0.99	+0.87	+0.12	+0.65	+0.78	-0.13
Franche-Comté	+1.04	+0.88	+0.16	+1.14	+0.88	+0.25	+0.96	+0.79	+0.17
Languedoc-Roussillon	+0.91	+0.24	+0.67	+1.57	+0.29	+1.25	+0.68	+0.14	+0.55
Limousin	-0.12	0.00	-0.12	+0.05	-0.11	+0.16	+0.05	-0.17	+0.22
Lorraine	+1.56	+1.24	+0.32	+0.58	+1.11	-0.52	+0.38	+0.79	-0.41
Midi-Pyrénées	+0.51	+0.29	+0.22	+1.02	+0.29	+0.73	+0.54	+0.17	+0.37
Nord	+0.97	+1.05	-0.07	+0.70	+0.92	-0.22	+0.37	+0.80	-0.55
Basse-Normandie	+0.44	+1.05	-0.61	+0.70	+0.90	-0.20	+0.53	+0.74	-0.21
Haute-Normandie	+1.19	+1.10	+0.09	+1.13	+0.99	+0.14	+0.93	+0.86	+0.07
Pays de la Loire	+0.72	+0.95	-0.23	+0.80	+0.88	-0.07	+1.01	+0.82	+0.18
Picardie	+0.84	+0.94	-0.10	+1.07	+0.86	+0.21	+0.91	+0.74	+0.17
Poitou-Charentes	+0.50	+0.74	-0.25	+0.37	+0.58	-0.22	+0.45	+0.44	+0.01
Provence-Côte-d'Azur	+2.16	+0.41	+1.75	+2.67	+0.49	+2.18	+1.59	+0.28	+1.30
Région Parisienne	+1.90	+0.72	+1.18	+1.46	+0.77	+0.69	+0.97	+0.79	+0.17
Rhône-Alpes	+1.35	+0.59	+0.76	+1.63	+0.73	+0.90	+1.10	+0.67	+0.43
Total FRANCE	+1.09	+0.69	+0.40	+1.14	+0.67	+0.47	+0.81	+0.58	+0.23

Source: Recensement Général de la Population, 1975

15

where there are fewer major cities. Eastern France, with its larger number of urban/industrial centres, affords a greater range of intervening opportunity to the migrant (Fielding, 1966).

The broad pattern of population change is evident from the table which shows that nine regions experienced a net migratory loss between 1954 and 1962. For the most part these were the more rural regions, especially those of the west and centre, where the young were leaving farms and villages for the cities. The greatest gains were recorded by the Paris Region and by Provence and Rhône-Alpes which included the fast-growing cities of Marseille, Lyon and Grenoble. Migrants were already leaving the old industrial towns of the Nord in the 1950s but the high rate of natural increase ensured an overall growth rate close to the national average. Account must also be taken of the arrival of a replacement population, especially of foreigners, to work in mining and other unattractive occupations. Lorraine, where the coal and steel industries were still expanding, experienced a net inflow of migrants between 1954 and 1962.

The 1960s saw the return to France of the <u>pieds noirs</u>, together with an increasing number of foreign workers, who helped to swell the population of the most urbanized regions of Paris, Provence-Côte d'Azur and Rhône-Alpes. They also contributed to growth in other parts of the Midi: Languedoc-Roussillon (+1.5% per year), Midi-Pyrénées (+1.02%). Rural loss continued, the removal of young people being most evident in the isolated upland areas where rates of natural increase were low and, in the case of Limousin, negative. Problems of contracting basic industries are evident in the migration of population from both Nord and Lorraine, a trend which continued throughout the succeeding period, 1968-75, in spite of government-assisted efforts to diversify the employment base in these regions.

All 21 regions experienced an increase of population between 1968 and 1975 but the fall in the birth rate and the reduction in immigration from overseas is evident in the lower overall growth. Numerically the largest increases continued to be in Paris (+613,000), Provence-Côte d'Azur (+366,000) and Rhône-Alpes (+358,000), but there were significant advances in Centre and Picardie, and also in three regions of the west (Bretagne, Pays de la Loire and, to a lesser extent, Poitou-Charentes), a consequence of the decentralization of employment that was taking place to the towns of the Paris Basin and even beyond. Within the Paris Region itself there was a considerable redistribution of population, the <u>ville</u> losing 290,000 whilst the four <u>départements</u> of the <u>grande couronne</u> experienced annual growth rates ranging from +2.8% to +4.6%.

It is apparent, even from figures relating to planning regions, that striking differences had emerged by the 1970s in the fortunes of cities in different parts of France. The

TABLE 5. ANNUAL RATES OF POPULATION CHANGE ACCORDING TO SIZE OF TOWN

Unités Urbaines	1954-1962 TOTAL	1962-1968			1968-1975		
		TOTAL	Natural Increase	Migration	TOTAL	Natural Increase	Migration
20,000	1.3	1.5	0.7	0.8	1.2	0.6	0.6
20 - 50,000	2.1	2.2	0.9	1.3	1.5	0.9	0.6
50 - 100,000	2.2	2.0	0.9	1.1	1.3	0.9	0.4
100 - 200,000	2.4	2.3	1.1	1.2	1.4	1.0	0.3
200,000 - 2 million	2.0	2.0	0.8	1.2	1.0	0.8	0.2
Paris (agglom.)	1.9	1.3	0.8	0.5	0.4	0.8	-0.4
Urban TOTAL	1.9	1.8	0.8	1.0	1.0	0.8	0.2
FRANCE	1.1	1.1	0.7	0.5	0.8	0.6	0.2

Source: Recensement Général de la Population, 1975

contrast is greatest between the old industrial centres, mainly in the north (e.g. Denain, Lens, Bruay, Béthune) and Lorraine (Thionville, Longwy, Forbach),with their often-unattractive environment and poor job prospects, and the buoyant urban centres of the Rhône basin (Grenoble, Chambéry, Chalon-sur-Saône), the south (Montpellier, Aix, Grasse) and the fringes of the Paris Basin (Orléans, Reims, Caen, Troyes, Tours). Migratory movement in France is no longer dominated by the pull of Paris; instead one must recognize the existence of a greater number of urban-orientated systems of varying size and attractiveness to migrants (Winchester, 1977).

Table 5 isolates the urban component of population change over the period from 1954 to 1975. It shows rates of growth in the towns and cities that were far in excess of those relating to the population as a whole. By the 1970s, however, these were falling, and the disparity between the urban and national rates of change was reduced. This is partly explained by the weighting given to the figures by the agglomération of Paris which was experiencing a net loss of population by migration, but is also due to the apparent slackening in the rate of migration to the cities. Overall, natural increase was four times as important a contribution to growth as migration by 1968-75. In seeking to interpret these trends, however, it is important to remember that the migration figures for 1962-68 were inflated by the heavy immigration from over-seas, and that by the 1970s there was increasing movement over city boundaries to nearby 'rural' communes. Averages also obscure inter-regional variations, out-migration exceeding natural increase in a few of the more depressed industrial towns.

When a distinction is made between groups of towns of different sizes, it is clear that the reduction in rates of growth has been greater in the large towns (over 200,000) than in the small and medium-sized ones; the highest rate of increase between 1968 and 1975 being in towns with a population of 20,000 to 50,000 (1.5% per year).

The importance of small towns in the French urban system is borne out by Table 6 which shows that no fewer than 8 million people lived in towns of under 20,000 population in 1975. Overall their population had increased by nearly 9% between 1968 and 1975, a rate of growth that was little less than that recorded by the medium-sized towns. The figures lend some support to the argument that the 1970s witnessed a reaction against the problems and pressures of life in the cities, but this does not necessarily imply a wide dispersal of population in favour of small towns throughout the country. Many of the fastest-growing of the small towns are those, in fact, which are within easy travel distance of the major cities.

The shift in the balance of growth from larger to smaller towns is nevertheless striking and it has had quite a profound effect on the rank ordering of towns in the lower half of the urban hierarchy. P.N.Jones (1978) has interpreted this in terms of the redistribution of economic activity and popu- lation that has been taking place in France, urban growth stimuli being spread more equitably over the country as a whole, thus bringing about changes in the national urban system. The inferences to be drawn from the growth of the medium-sized and small towns are discussed more fully in Chapter 4.

TABLE 6. POPULATION CHANGE BY SIZE-CATEGORY OF TOWN, 1968-75

Unités Urbaines	1968 (thousands)	1975 (thousands)	Change 1968-75 %
20,000	7,339	7,987	8.83
20 - 100,000	6,882	7,597	10.39
100 - 200,000	4,050	4,449	9.85
200,000 - 2 million	8,291	8,887	7.19
Paris	8,195	8,424	2.79

Source: Recensement Général de la Population, 1975

Analysis of the Urban System
        Study of the urban system involves consideration of the size, spacing and hierarchical relationships of towns and cities at both the national and the regional scale. The cities that make up the system are functionally interdependent and interaction between them takes many forms. The most obvious expression of such inter-relatedness lies in the regular movement of people and goods, utilizing road and rail and, in the case of the largest towns, air services. But contacts are also maintained by telephone and, increasingly, by still more sophisticated forms of communication. The transmission of information by such media is less easy to measure than the transport of passengers or freight, but is no less important to an understanding of how the system works. Indeed, the nature of information flows within the 'quaternary' sector of high-order services is essential to an appreciation of what goes on within and between the cities of the modern 'post-industrial' urban system (Pred, 1973). Hierarchical relationships are evident in the structural organization of industry and commerce and lead to a consi- deration of the regional function of cities. Central place theory has provided an idealized explanation of the size and spacing of market centres at the regional scale, but there are many other circumstances that account for the actual spatial arrangement of towns, as demonstrated for example by Lösch.

Studies of the urban system have become more sophisti-
cated with time.   Early examples were preoccupied with population totals and with the employment structure of towns and
cities.   The result was rarely more than a classification of
urban centres, little attention being paid to the nature of
functional linkages that bound those centres together in a
single system.   Later work has shown more concern for such
ties, and has also sought to investigate the socio-economic
make-up of towns in greater depth.   This has involved consideration of the division of labour and type of work, the
presence or otherwise of particular sectors of industry, and
the distribution of wages and skills.   More account is thus
taken of the social structure of cities (Aydalot, 1976), and
the hierarchical arrangement of urban centres is related to
the division of work that characterizes the modern industrial
corporation (Simmie, 1981).   Studies that have been made of
the French urban system over the last 30 years reflect this
change in approach as the examples below seek to demonstrate.
    A final word concerns terminology.   The phrase 'système
urbain' has now been adopted by French writers, but until
recently the terms 'trame' (web), 'réseau' (network) or
'armature' (framework) were more commonly employed.   They
seem to have been used inter-changeably, with no subtle distinction implied by the choice of word.

The Functional Classifications
    Early attempts at classification were made in the 1950s
by George (1952) and Gravier (1958).   Each distinguished five
categories of urban centre below the level of the capital,
noting the predominance of small towns in France and the
limited number of large ones.

    George (1952)   le centre local (e.g.Luçon)
                     la capitale régionale (Angers)
                     la ville marchande industrialisée (Le Mans)
                     la ville industrielle (Saint-Etienne)
                     les grandes agglomérations (Marseille)

    Gravier (1958)   le petit centre (e.g. Sélestat)
                     la ville moyenne (Aurillac)
                     la capitale secondaire or
                               'ville-relais' (Brest)
                     la capitale régionale (Dijon)
                     la métropole internationale (Lyon)

    Gravier's classification is interesting for the use of
such terms as ville moyenne and ville-relais that have since
become current in urban planning circles.   His paper was
followed by the lengthier study of Coppolani (1959) who also
recognized a five-tier hierarchy (excluding Paris) in his

réseau urbain, but expressed doubts about a possible sixth
category of village-centre (chef-lieu communale) that in some
circumstances could be regarded as an 'embryonic town'.
Exploring this problem he made the useful distinction between
the large Mediterranean village which probably had a population
in excess of 2,000 but which was predominantly 'rural' in its
occupational structure, and the much smaller 'bourg', found
in areas of predominantly dispersed settlement such as
Brittany and Limousin, where the population of small traders,
teacher and parson, were more exclusively 'tertiary' in
employment.   It was to the bourg that he applied the term
'embryonic town' on account of its service role to the surrounding region.

    Coppolani (1959)   (le chef-lieu communale)
                              la bourgade
                              le centre local
                              la ville maîtresse
                              la sous-capitale
                              la capitale régionale

    Coppolani's work drew heavily on the ideas of Christaller
and the contemporary classifications of such British scholars
as Smailes.   His bourgade, 'neither village nor town', had a
population often of no more than 1,000-2,000.   Its urban
status derived from its ancient market function, possibly
dating from the Middle Ages, and its possession of non-food
shops such as chemist or haberdasher and of services like
those of a dentist, insurance agent or agricultural advisor.
The bourgade is often the chef-lieu of a canton and as such
has a limited administrative function with offices of the
gendarmerie and tax collector.   Its sphere of influence
extended for some 8 to 10 km serving, in the 1950s, an agricultural population of some 5,000-8,000.
    The centre local was described as having a population of
around 5,000 but this could reach 10,000 when it included a
manufacturing (Mazamet) or tourist (Aix-les-Bains) function.
Its market might deal in specialist agricultural products,
e.g. veal or flowers produced in the surrounding area, and its
shops would be similar to those of the bourgade, possibly
including furniture stores and a perfumery.   Services were
likely to include a weekly paper, cottage hospital and,
administratively, the town might have the status of sous-
préfecture.   Local bus services extended its sphere of
influence to 30-35 km.
    Coppolani's ville maîtresse, which he also described as
chef-lieu de pays, ranged in size from 10,000 to over 100,000,
but was typically in the range 30,000-50,000 and was often
the location of a préfecture, e.g. Agen, Annecy or Chartres.
Some départements possessed more than one such town, however

(Mâcon, Autun and Chalon-sur-Saône), and the category included
a number of industrial towns and ports in the north.   In all,
the number of villes maîtresses was about 120.   Their services
included banks, wholesaling and a Chamber of Commerce, a
public library, theatre, and a hospital offering specialist
medical treatment.

The distinction which Coppolani draws between capitale
régionale and sous-capitale is not absolutely clear; there
is, for example, a category of capitale régionale contestée.
Eight cities, however, emerge as uncontested regional capitals:
Marseille, Lyon, Bordeaux, Lille, Rouen, Strasbourg, Toulouse
and Nantes.   All had populations in 1954 in excess of 200,000
and high order services.   Four others (Limoges, Clermont-
Ferrand, Nancy, Dijon), all of them former provincial capitals,
are placed in a sub-category of regional capital on account of
the 'relais' function they perform; in effect, filling in gaps
in the national space.   The group of 18 towns to which he
gives the title of sous-capitale - equivalent to Gravier's
capitale secondaire - have some of the historical, economic
and institutional attributes of the regional capitals, but
have a more limited sphere of influence, depending on the
latter for certain higher order services.   They include
Grenoble, Nice, Montpellier, Pau and Brest, and also towns
which are within the orbit of Paris such as Amiens, Orléans,
Bourges, Tours and Le Mans.

Coppolani's scheme, largely descriptive in nature and
based primarily on the established 'central place' role of
French towns, does not differ greatly from the system worked
out by Hautreux and Rochefort (below) which has had such a
strong influence on French planning.   Seven of Coppolani's
eight regional capitals - Rouen is the exception - were later
to become the métropoles d'équilibre and his concept of the
ville-relais has enjoyed wide currency.

Although it was published in 1963, the study of the
French urban system carried out by Carrière and Pinchemel
was based largely on data drawn from the 1954 census.   But
despite its age, the book merits attention both on account of
the detailed nature of the analysis carried out, particularly
on the employment structure of French towns, and for its
general comments on the nature of the urban system as it was
at that time.

Stress is laid on the uneven distribution of towns and
cities making up the system, 68% of the population living in
only 23 départements.   Except for Paris, these are mostly
found towards the periphery of the national space, to the
north, east and south-east.   The contrast between the eastern
and western halves of the country is also striking.   The east
is characterized by a number of densely-populated urban
regions separated by the sparsely-peopled 'deserts' of the
southern Alps, the south-east of the Paris Basin, and the

south-eastern portions of the Massif Central.   The west, by
contrast, has a more regular distribution of small towns with
few heavily urbanized areas and no very extensive 'deserts'.
Thinly-populated districts are relatively limited in extent,
e.g. parts of interior Brittany, the hills of Perche and
portions of the Norman bocage.   At the regional level there
are equally marked contrasts between those départements or
groups of départements that are dominated by a single city
(Toulouse, Bordeaux, Dijon, Clermont-Ferrand, Limoges) and
those, like Oise, with numerous small urban centres, or the
industrial areas where functionally-specialized towns do not
fully 'urbanize' the region.   The French urban system thus
exhibits 'urbanisation lacunaire et discontinue ... avec des
hypertrophies et des lacunes'.

TABLE 7. PROPORTION OF THE URBAN POPULATION LIVING IN DIFFERENT
         SIZE-CATEGORIES OF TOWN, 1954 (after Carrière and
         Pinchemel)

Unités Urbaines (1954 definition)	Size-category	Number	Percentage of the urban population
	5-20,000	612	22.6
	20-50,000	117	14.4
	50-100,000	29	7.8
	100-200,000	17	8.9
	200,000-1 million	10	15.9
	Paris	1	19.5

Carrière and Pinchemel also contrast the large number of
small towns in France with the under-representation of those
in the range 50,000-200,000 which accounted for only 16.7% of
the urban population in 1954 (Table 7. N.B. the definition of
'unité urbaine' used in 1954 over-inflates the number of
small towns compared with later definitions, but in no way
invalidates the conclusions drawn).   Their comments on one
category of medium-sized town are significant in view of later
planning policies: 'la catégorie de taille 50,000-100,000
habitants a présenté certains signes pathologiques: sous-
représentation numérique, faiblesse des fonctions spécifiques,
rôle mal défini entre les deux catégories de tailles voisines'.
Their more general description of the French urban system is
also worth quoting in full as background to the formulation of
national planning strategies that was taking place at this
time.

La France, déjà faiblement urbanisée ne possède pas un

réseau urbain organique; elle est composée d'une série de
réseaux urbains régionaux discontinus; entre ces réseaux,
ayant chacun sa structure propre, correspondant à des
pays d'ancienne tradition urbaine, existent de larges
zones inurbanisées ou mal urbanisées, par un ensemble
inorganique de petites villes, ou par une seule grande
ville qui émerge - isolée - du monde rural environnant.
(p.307)

The functional classification of Carrière and Pinchemel
allocated towns to one of three categories based on the proportion of the working population engaged in particular
activities.   These were:'villes mono-industrielles', 'villes
poly-industrielles', and 'villes de services', but allowance
was also made for degree of specialization.   Three-quarters
of French towns of more than 20,000 population fell in one or
other of the sub-categories of 'ville de service' (50% of their
working population in the tertiary sector).   The spatial
pattern is thus one in which service towns predominate, but
one which also displays considerable diversity at the regional
scale owing to the patchy intrusion of towns with varying
degrees of industrial specialization.
    Not dissimilar from the work of Carrière and Pinchemel is
the study carried out by Prost (1965) which explored the
relationship between town size, administrative status and the
presence of various kinds of services.   This, too, was largely
based on results of the 1954 census, however, and a full description would not seem justified.   But Prost's work is also
interesting for the attention paid to the sphere of influence
of the service centre and the link between le rayonnement
urbain and a town's position in the hierarchy.   Her research
was limited, however, to the sphere of influence of Lyon and
of the smaller towns of the département of Rhône.   Others
sought to apply the same ideas to the whole of France.

Spatial and Multi-variate Approaches
    The early 1960s saw an increasing awareness in France of
the dynamic role played by the city in the life of its surrounding region and of the possibilities of using this spatial
relationship to spread economic benefits throughout the region.
Not that the idea was new.   As early as 1913, Vidal de la Blache
had observed that 'aujourd'hui, c'est la ville qui crée la
région ... c'est à l'attraction qu'elle exerce autour d'elle
que se mesure l'étendue de la région qui doit lui être
attribuée' (in Bloch, 1913).   Other writers echoed the same
idea, but widespread public interest had to await publication
of Perroux's (1950) theories of the growth pole and more
particularly of Boudeville's translation of these fundamen-
tally economic concepts into a geographical or spatial context
(e.g. Boudeville, 1958).

Having accepted the relationship between city and region, it becomes necessary to measure the extent of this sphere of influence. Piatier (1956) devised a method of measuring the trade area of a town and this has been widely used in empirical work, first in relation to a number of départements in the south-west, subsequently in other parts of France. Vivian's (1959) study of Grenoble is a detailed analysis of the zone of attraction of a single town, and Chabot (1960) published a map of spheres of influence of the principal French cities at a scale of 1:625,000. Of more profound importance so far as subsequent policy-making was concerned, however, was the work of Hautreux and Rochefort.

An early paper by Rochefort (1957), which explored methods of studying the urban system, was followed by Hautreux's (1972) delimitation of the regional spheres of influence ('ressorts d'influence') of the larger French cities. Measurement was based on three criteria - telephone calls, the movement of rail passengers, population migration - and the results were accompanied by a map. Nine cities (Paris apart) were described as having 'une attraction de caractère régional': Nancy, Lille, Strasbourg, Lyon, Marseille, Montpellier, Toulouse, Bordeaux and Nantes, although Montpellier was recognized as looking to Marseille for certain high order services. Seven other places were defined as 'villes attractives secondaires': Clermont-Ferrand, Limoges, Rennes, Dijon and Tours - whose role was limited by the influence of Paris - Grenoble and Saint-Etienne, which were similarly overshadowed by Lyon. It was also suggested that this hierarchy of spheres of influence was paralleled by one based on the nature of the functions exercised by the cities and also by their importance as centres of population.

In 1962 Hautreux and Rochefort undertook a study of what was described as 'l'armature urbaine française' for the Ministry of Construction's Centre d'Etudes Economiques et Sociales. The results of this enquiry were issued in the form of two reports by the Ministry in 1963 and 1964 (Hautreux and Rochefort, 1963, 1964). According to the authors, 'On conviendra d'appeler 'armature urbaine' du pays l'ensemble hiérarchisé de ces centres, qui en assurent l'encadrement tertiaire, considérés dans leur localisation à travers l'espace national et dans les découpages de cet espace qui résultent de leurs zones d'influence ... la disposition des centres inter-médiaires et locaux au sein des régions majeures par les zones d'influence des centres régionaux correspond à la notion de réseau urbain régional' (Hautreux and Rochefort, 1965).

On the basis of various criteria, the study attempted both to rank the leading urban centres and to delimit their spheres of influence. In all, 208 towns and cities were examined. Ranking was carried out on the basis of four sets of data relating to, (i) population (total in 1962, division

TABLE 8. LEADING CENTRES OF THE URBAN HIERARCHY ACCORDING TO HAUTREUX AND ROCHEFORT

Urban Centre	Scores based on the four criteria of				Total Scores
	Population	'Economic' Services	'Social' Services	External Influence	
Métropoles Régionales					
Lyon	65	179	179	75	498
Marseille	65	180	176.5	75	496.5
Bordeaux	65	180	172	75	492
Lille-Roubaix-Tourcoing	65	176.5	168.5	75	485
Toulouse	65	169.7	175	75	484.7
Strasbourg	64.3	176.5	171	70.2	482
Nantes	65	172	159.5	75	471.5
Nancy	61.35	161.5	158.5	71.4	452.75
Centres Régionaux					
Grenoble	59.1	153.5	158	57.2	427.8
Rennes	53.2	145.4	159.5	61	419.1
Nice	62.9	148.9	152	49.8	413.6
Clermont-Ferrand	57.65	136.5	155	61.6	410.75
Rouen	65	138.5	153	54	410.5
Dijon	53.9	131.2	164.5	59.2	408.8
Montpellier	47.45	133.2	160	56	396.65
Saint-Etienne	63.6	131.5	127	53.8	395.9
Caen	44.25	148.8	131.5	61	391.55
Limoges	53.3	130	139	58.6	380.9

Villes à Fonction
Régionale Incomplète

Metz	51.45	126.5	130.5	59	367.45
Tours	54.95	108.8	135.5	58	357.25
Amiens	49.4	124.3	125.5	53.6	352.8
Le Mans	52.6	114.5	127	53	347.1
Orléans	50.6	113.2	139	44.2	347
Angers	51.55	110.3	137	46	344.85
Reims	54.6	118	122.5	43.8	338.9
Besançon	44.2	102	136.5	51.2	333.9
Nîmes	46	109.3	123	53	331.3
Mulhouse	55.45	122	98.5	41	316.95
Troyes	48.55	102.8	111	48.2	310.55
Le Havre	59.8	112.4	95	38.2	305.4
Perpignan	41.4	105.3	114.5	42.2	303.4
Poitiers	37.85	78.5	134	50.4	300.75
Bourges	39.95	92.4	119	48	299.35
Avignon	41.4	111	96	40.4	288.8
Angoulême	40.2	97.8	105	44.6	287.6
Bayonne-Biarritz	49.7	101.5	94.5	40	285.7
Brest	50.5	98.1	95	42	285.6
Valence	36.9	90.8	102	53.6	283.3
Annecy	35.1	97	102.5	48.6	283.2
Pau	40.8	93.5	107	39.8	281.1
Chambéry	33.35	101.2	99.5	46.8	280.85
Toulon	57.75	84	92	44	277.75

between secondary and tertiary occupations), (ii) services associated with the economy (wholesaling, offices, banking, consultancy and trade organizations, air links), (iii) services of a social and cultural nature (higher education, administration, the arts and sport, specialist medical services), and (iv) external influence (population of the region, road and rail services). Scores were allocated to individual centres on the basis of the importance of these various criteria and a classification produced (Table 8).

The ranking exercise yielded 8 métropoles régionales, 10 centres régionaux and 24 towns that were described as villes à fonction régionale incomplète (what Noin (1976) calls centres sous-régionaux). The remainder of the 208 towns, not listed in Table 8, include 32 places, many of them chefs-lieux of départements, which exercise a regional function that is largely confined to that administrative unit; the rest serve a more local area. Within the leading category, Hautreux and Rochefort drew attention to the somewhat weaker position of Nantes and Nancy compared with the rest. Otherwise the eight are said to constitute a reasonably coherent group of centres offering a high order level of services. The title of métropole régionale is adopted since it appears to be standard usage, but the authors note that the dominance of Paris is such that these cities compare unfavourably for the most part with cities in other West European countries to which the same term is applied.

There is a significant break in the table of scores between Nancy, the least well-equipped of the métropoles, and Grenoble, which heads the list of centres régionaux. The latter offer a more or less complete range of services, though not always of the highest order and for a more restricted hinterland. They include Rouen and Nice, large but with a truncated service area, and Saint-Etienne, which scores badly on its provision of 'social' services, perhaps on account of its industrial history and proximity to Lyon. Contrast, in this respect, Montpellier, with its advanced medical services and large university.

Variety is still more apparent in the list of 24 towns offering an incomplete range of regional services. They are all places that score well in some respects, but are deficient in others, and have a restricted hinterland compared with the regional centres. This is especially true of towns in the outer parts of the Paris Basin. Industrial towns tend to score badly - Le Havre, Toulon - with notable absentees from the list such as Valenciennes, Douai and Saint-Nazaire. Poitiers, with a very limited industrial base is, by contrast, promoted in the list by virtue of its service role.

Figure 1 shows the location of all 42 of the centres listed in the Table and the sphere of influence of the métropoles and centres régionaux delimited by Hautreux and

Figure 1. The French urban system according to Hautreux and Rochefort

Rochefort. The most striking feature of the map is the peripheral location of all the métropoles and the absence, apart from Lille in the industrialized north, of any such métropoles within 200-300 km of Paris. Characteristic rather of this Parisian aureole are towns defined as having an incomplete regional function. Elsewhere the distribution is irregular, with towns such as Perpignan and Pau filling in obvious 'gaps' in the national territory. Only Rhône-Alpes amongst the régions de programme appears to have anything approaching the theoretical hierarchy of urban centres, with a system of towns that perform a complementary range of functions.

The work of Hautreux and Rochefort was influential in the choice by the government of cities to serve as métropoles d'équilibre. Before examining this strategy, attention may be drawn to two studies of the urban system carried out in the 1970s - those of Noin (1976) and of Pumain and Saint-Julien (1978).

The Changing System

Noin's work is of interest, both for its detailed study of the urban hierarchy of a single region - Basse-Normandie - and for its observations on the national urban system. The latter are based, firstly on a ranking exercise involving 249 centres having at least 600 persons employed in commerce and private services (in 1968) and, secondly, on a series of maps showing the spheres of influence of French towns based on (i) commercial activity (after Piatier), (ii) catchment areas of the universities, (iii) telephone calls, and (iv) public transport services. Great emphasis is laid in the conclusions on the over-riding importance of Paris with regard to high order activity in France. Noin also comments on the disadvantages to residents in the isolated 'campagnes profondes' of distance from towns offering advanced medical or educational services.

He proceeds to suggest, as an alternative to the existing planning regions, a different regional pattern based on the spheres of influence of the leading urban centres. These regions (15), which group départements as do the planning regions, are based on the two highest orders of an urban hierarchy that is almost identical to that of Hautreux and Rochefort. At the top of the hierarchy are the eight métropoles, which Noin describes as 'capitales régionales'. Below them is a category of 'ville pouvant être assimilée à une capitale régionale', but it includes only seven of the ten centres régionaux of Hautreux and Rochefort, Rouen, Caen and Saint-Etienne being demoted on account of the effect on their service function of either Paris or Lyon. These three join the next group of 24 in a class of 'ville ayant un rôle régional plus réduit'. Significantly, twelve of these appear

within Noin's vast Paris region which embraces 32 <u>départements</u>.
Then comes the 'agglomération importante mais sans fonction
régionale bien nette', including many of the departmental
<u>chefs-lieux</u> and larger industrial towns.

The principal value of the detailed study carried out by
Pumain and Saint-Julien lies in its emphasis on what has been
happening to the urban system over a critical period of change
in France, the 20 years before 1975. For the purposes of
their analysis, the authors draw on data from the censuses of
1954, 1962 and 1968 relating to 138 <u>agglomérations</u> which had
a population of at least 20,000 in 1954, and on the census of
1975 for 99 <u>agglomérations</u> of 50,000 or more. Most use is
made of employment data, but the study also takes account of
income levels, social categories, wealth and demographic change.
This enables the authors to classify towns on a more sophisti-
cated basis than that achieved by Carrière and Pinchemel, and
to make certain predictions concerning future trends.

Several classifications are, in fact, presented. The
first employs a similar method to that of Florence (1948),
comparing the actual distribution of employment against an
expected distribution, and achieves 18 classes of town in
three broad groups: those having a high degree of industrial
specialization (28), others that display specialization in
industry or services (71), and ones which are relatively
unspecialized (39). A further classification is derived from
observation of growth trends in the various employment groups
and, following this, a third one makes use of two variables
which the authors describe as 'image de marque' (utilizing data
on economic activity, social type, standard of living, population change) and 'axe de modernité' (income levels, kind of
work, employment growth). Figure 2 is based on this multi-
variate analysis of Pumain and Saint-Julien and refers to 88
<u>agglomérations</u>, each with a population exceeding 50,000 in
1975.

A number of conclusions emerge clearly from the study:

  (i)   that changes in the balance of employment (the pro-
       portion of the towns' working population engaged in
       'secondary' occupations rose from 45.3% in 1954 to
       46.9% in 1962, subsequently falling to 44.6% in 1968
       and 41.0% in 1975) have taken place without any major
       adjustments to the urban system,
         ' ... ces vingt dernières (1954-75) sont, en dépit
         des apparences, caractérisées par une assez grande
         stabilité des composantes d'activité des structures
         urbaines.'

  (ii) that, except for the larger urban centres, no clear
       relationship is observable between city size and
       functional classification,

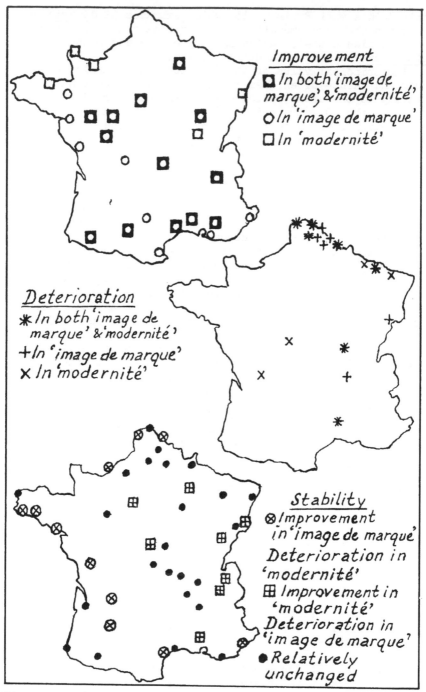

Figure 2. Change in the urban system since 1954, after Pumain and Saint-Julien

(iii)   that (notwithstanding (i) above) activities are more
equally represented in French towns than was the case
in 1954, i.e. geographical concentration of activity has
been reduced,

>'Quelques modifications d'ensemble indiquent
homogénéisation des profils d'activité des villes,
élargissement du tronc commun d'activité, réduction
des plus fortes spécialisations, apparition de
spécialisations nouvelles dans les villes au profil
très moyen.'

(iv)   but that the creation of new jobs that require low
qualifications and are relatively poorly paid may have
<u>increased</u> social disparities.

>'Elle pourrait en outre renforcer l'hypothèse d'une
spécialisation sociale croissante des villes en
rapport avec les formes nouvelles de la division
du travail.'

This final conclusion gives added weight to what may be
observed in Figure 2, implying as it does that job diversifi-
cation under government programmes may have conferred fewer
benefits on such towns as Bruay, Thionville, Lens or Longwy
than is sometimes claimed.

    Using what is admittedly a very coarse grid, the seven
<u>zones d'étude et d'aménagement du territoire</u> (ZEAT), the
authors observe an increased homogeneity in employment structure at the regional scale. Only the north, with its greater
emphasis on industry, and the south-east with its corresponding weakness in manufacturing, stand apart from this model of
the region as a microcosm of the whole. Growing inter-
regional similarity should not, however, be allowed to obscure
a high degree of intra-regional diversity.

    Pumain and Saint-Julien emphasize the structural conver-
gence in economic activity that has taken place within the
French urban system. It is suggested that this may have been
associated with the diffusion of industrial activity that took
place in the last cycle (up to 1962), borrowing the idea of
the product-cycle from Aydalot (1976). Diffusion theory
would also help to account for differences between, for example,
the north-west and the south-west in the ability of their
towns to acquire new employment. The deconcentration wave
from Paris is claimed to have reached the outer parts of the
Paris Basin by 1968, Alsace and the northern edge of
Aquitaine by 1975. Finally, account must be taken of postwar
indicative planning. The suggestion made by the authors that
industrialization has shown a tendency to diffuse down the
urban hierarchy may be related to strategies that put the
emphasis, first on the larger <u>métropoles</u>, and later on the
medium- and small-sized towns. This theme is explored in the
two succeeding chapters.

Chapter 3.

PLANNING THE URBAN HIERARCHY (I)
THE METROPOLES D'EQUILIBRE

The rise of Paris to a position of unchallenged supremacy within the French urban system was noted in the previous chapter. By 1962, Paris was some eight times larger than its nearest provincial rival, Lyon, and examination of the rank-size distribution of the 30 largest provincial cities shows most of them with populations far below the 'ideal' (Figure 3).

But population figures alone provide an inadequate measure of the extent to which Paris had come to dominate the political, economic and cultural life of the country. The absence of an administrative hierarchy between the national capital and the préfecture had the effect of concentrating decision-making in Paris. More than 90 of the 100 largest French companies had their headquarters in Paris, and 91% of what Labasse (1974) calls 'la puissance financière' was likewise to be found in the capital. Hautreux and Rochefort (1965) refer to the 'trains d'affaires' which enable the provincial businessman to reach Paris and return within the day, having negotiated his loan, or obtained permission for whatever development is proposed. The intellectual pre-eminence of Paris was reduced somewhat in the 1960s as attempts were made to decentralize higher education, but Paris still retained two-fifths of the students in the grandes écoles and some 60% of scientific research was carried out there (Noin, 1976).

Early postwar planning showed an awareness of the sort of problems Gravier had highlighted in Paris et le Désert Français (1947), but solutions sought were of a broad-brush nature. They included policies aimed at bringing about the decentralization of industry from Paris (the Petit Plan of 1950) and the complementary strategy of promoting regional development by means of comités d'expansion économique, set up in 1954. But it was not until 1962 and the formulation of the Fourth National Plan (1962-66), that any significant attempt was made to manipulate the urban system as part of a coherent strategy of national development. Crucial to this new policy was the

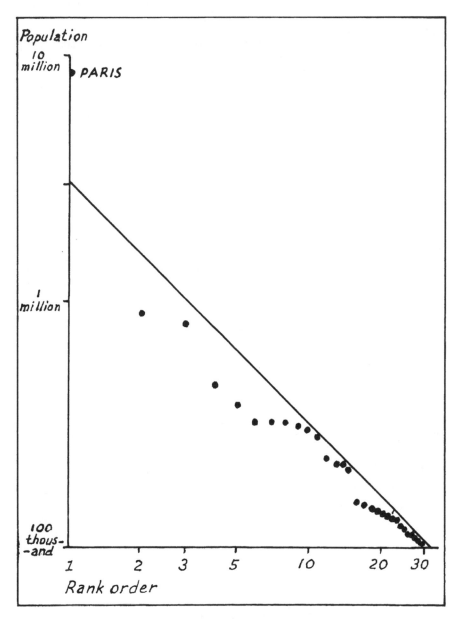

Figure 3. Rank-size distribution of the 30 largest French
           towns, 1962

establishment in 1963 of DATAR (Délégation à l'Aménagement du Territoire et à l'Action Régionale), attached to the Prime Minister's office. DATAR's responsibility lay in the co-ordination of planning schemes proposed by the various ministries and departments, and in ensuring that a regional component was built into the National Plan.

The 22 French planning regions date, in an embryonic form, from 1955 when committees of regional economic expansion were set up. The role of what came to be known as régions de programme and, later, circonscriptions d'action régionale, was subsequently strengthened in the 'reforms' of 1964 when the office of préfet de région was established. The préfet's responsibility was to make the needs of the region known to central government and he was assisted in this task by a CODER (Commission de développement économique régional) of appointed experts drawn from industry, the unions, universities, etc., together with a number of representatives of local government. The préfet de région was the préfet of the most populous département within the region, and the presence of the préfet de région conferred an additional administrative importance on what also tended to be the largest towns (Metz in Lorraine was an exception). This extra status was slight, however, in comparison with the distinction accorded to those towns chosen to act as métropoles d'équilibre.

The policy of the 'balancing metropolis', adopted in 1964 and actively promoted by DATAR, was a response to the over-whelming predominance of Paris, and recognized the very weak role of the French provincial capital in comparison with its counterpart in other West European countries. Help was urgently needed if the pull of Paris was to be offset and it was felt that only by advancing the larger provincial towns could there be any prospect of doing this. Parisian com-panies, it was thought, might be persuaded to move a part at least of their operations to the provinces if the infrastruc-ture of these major cities were made more attractive. Once there, of course, there was always the possibility, according to the growth pole theorists, that benefits would spread throughout the region. '... la première nécessité étant de créer des pôles dotés d'une gamme complète et puissante de services auxquels toute une vaste région ait recours dans les domains les plus spécialisés ou pour obtenir les décisions les plus importantes' (Hautreux and Rochefort, 1965, p.676). Comparisons were made with the Greek city states, which claimed the title of métropole when they had founded urban colonies. The politique des métropoles d'équilibre also reflected France's involvement with the European Economic Community. With this came a desire to put French provincial cities on an equal footing with those of other Common Market countries, able to attract foreign investment and of sufficient importance to be drawn into the evolving European network of transport and communication services.

Eight cities, or groups of cities, were designated métropoles d'équilibre when the French parliament gave its approval to the strategy in 1964: Lille-Roubaix-Tourcoing, Nancy-Metz-Thionville, Strasbourg, Lyon-Saint-Etienne-Grenoble, Marseille-Aix, Toulouse, Bordeaux, and Nantes-Saint-Nazaire (Figure 4). These were the eight cities revealed by Hautreux and Rochefort's studies as having the most developed infrastructure, service provision and external contacts and thus being most likely to respond to the measures proposed. The latter were wide-ranging and included plans to promote new forms of employment, to raise the threshold of decision-taking in the service sector, to improve transport and communications links, and to carry out schemes of urban renewal and extension. (They are set out fully in a special edition of Urbanisme, No.89, 1965.) The programme, it was intended, would be completed by 1985.

In all but three of the eight cases, the métropole was linked with other towns in a bi-polar or tri-nodal group. This recognized the close physical links that existed between, for example, Lille, Roubaix and Tourcoing, or between Nantes and Saint-Nazaire. It also acknowledged functional associations as between Aix, with its university and law courts, and Marseille. The grouping of Saint-Etienne and Grenoble with Lyon is less easy to justify except in terms of the more developed urban system of Rhône-Alpes and the pattern of contacts implied by that. The link between Metz and Nancy may be regarded largely as a concession to the long-standing rivalry between these two urban centres, although it has some logic in the context of movement and exchanges taking place along the Moselle valley. It was supposed that, by grouping towns in this way, there would be greater opportunity of co-ordinating investment, particularly in infrastructure, and so of conferring benefits on the region as a whole.

The same principle is evident in the decision taken in 1966 to establish OREAMs (organisations d'études d'aménagement des aires métropolitaines), urban planning regions which enabled the developments proposed for the métropoles to be examined in a wider context. In particular, it was intended that their plans would make provision for the land needs of the cities where major schemes were to be implemented for new towns, motorways, airports or commercial centres outside the existing municipal boundaries. Metropolitan planning regions for this purpose were set up to assist five of the métropoles (Strasbourg, Bordeaux and Toulouse were the exceptions). In addition, several of the cities (Lille-Roubaix-Tourcoing, Strasbourg, Lyon, Bordeaux) took advantage of other legislation to negotiate with neighbouring communes and establish, from 1967, communautés urbaines. Less extensive than the OREAMs, these recognized the anomalous nature of some city

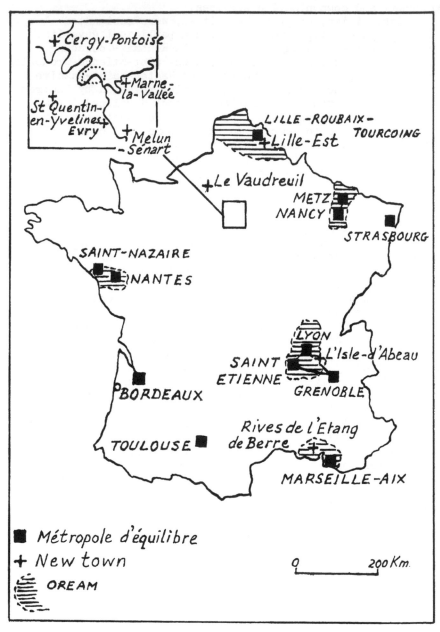

Figure 4. Metropoles d'équilibre and new towns

boundaries and the need to provide services and to plan for
the overspill as well as the city's own territory and population.   Reluctance to cooperate on the part of suburban
communes could be, and indeed in these cases was, overruled by
central government.

Lyon - Saint-Etienne - Grenoble

Greater Lyon has a population that now exceeds 1.5 million
and this total is expected to pass 2 millions by the end of
the century.   Though far behind Paris, Lyon holds an
unchallenged position as second business and commercial centre
of France.   The city lies at the heart of the growth region
of Rhône-Alpes, and it has been claimed that Lyon is the only
provincial city that is sufficiently large and well-equipped
to really deserve the title of balancing metropolis and to
succeed as counterweight to the dominance of Paris.   It has
also acquired an international role, whilst the urban environment has been transformed over the past 25 years by schemes
involving both new development and urban renewal.   The
extent of the change has been far in excess of that experienced
by any British provincial city.

On 22 September, 1981, President Mitterrand formally
inaugurated the new rail service between Paris and Lyon by
train à grande vitesse.   The TGV completes the 426 km
journey in 2 hours 40 minutes, and this will be reduced to
two hours when the whole of the new line which carries the
train has been completed towards the end of 1983.   As part of
its investment, the SNCF has also modernized Lyon's Perrache
railway station, opened a new station at la Part-Dieu, and
transferred to Lyon the 600 jobs involved in its supplies
department.   Approved by the government in 1974, work on the
TGV was begun towards the end of 1976.   Its opening symbo-
lizes for many the progress that has been made by the city of
Lyon over the past quarter of a century, though arguably just
as symbolic was the entry into service of the new international
airport of Satolas in 1975 and the inauguration of the first
section of the métro in 1978.

From Roman times to the present, the city of Lyon has
derived great benefit from its crossroads location, and it is
improvement to the transport infrastructure, utilizing this
natural advantage, that has contributed most to the city's
postwar transformation.   Lyon is now at the hub of a motorway
network, the backbone of which is the Paris-Marseille autoroute
opened in 1970.   The northern and southern sections of this
Rhône axis are joined by a tunnel under the Fourvière massif
completed in 1971, new bridges being built to carry the motorway across the Saône and the Rhône where it passes through the
city.   During the 1970s the network was gradually extended,
notably in the Alps where international road traffic makes
use of the road tunnels opened at Mont Blanc in 1965 and

Fréjus in 1980.   The long-awaited motorway link with Geneva and the Swiss network is expected in 1983.

Adding to the advantages conferred on the city by its road connections are the benefits that have followed from investment in other forms of transport.   The opening of Satolas airport, for example, and the subsequent extension of international air services, is a major factor in plans to re-establish Lyon as an international business and commercial centre.   Several oil and gas pipelines pass through Lyon, supplying both fuel and a source of raw materials for the Feyzin refinery (1965) which has a big part to play in the city's chemical industry.   Finally, reference might be made to waterways, the early 1980s having seen completion of the project to make the Rhône navigable for 'European' barges between Lyon and the sea at Fos.   Large barges can also proceed upstream to the Saône, making use of the Pierre-Bénite lock and barrage opened in 1967, but little progress has been made with the much-debated deepwater link with the Rhine. Completion of this artery would confirm what Prime Minister Barre described as the 'vocation européenne' of Lyon and its region (Le Monde, 29 January, 1979).

In the sixteenth century, Lyon was one of the most important banking centres of Europe, but centralization resulted in this lead being lost to Paris and even Crédit Lyonnais, founded in 1863, found it necessary to transfer its headquarters to the capital.   One of the objects of the politique des métropoles d'équilibre has been to restore this financial reputation, hopefully leading to a situation where Lyon might rival such established centres of finance as Zurich, Milan or Frankfurt.   Attempts have also been made to tempt both French and foreign companies to open offices in the city, a special kind of moral pressure being reserved for companies such as Crédit Lyonnais, Rhône-Poulenc and Pechiney which had their origins in the Lyon region.

The policy has had some success, though less than its more optimistic protagonists intended.   An impressive number of banks, including several American ones, have opened branches in Lyon since the early 1960s, and local managers now have greater discretion to grant short-term credits. There has also been a marked increase in the amount of insurance transacted in the city.   Many company offices have been opened, but most of these are of subsidiaries and not company headquarters.   Fairly typical has been Rhône-Poulenc's decision to move just its textile section to Lyon.   A number of foreign companies have followed the example of Richier-Ford and Black and Decker and chosen Lyon as the headquarters of their French or, in the case of the former, European operations.

Closely associated with the above has been the promotion of scientific research, a matter over which the government is able to exercise rather greater control.   Lyon's reputation as a centre for medical research has been enhanced, for example, by the presence since 1966 of the International Cancer Research Institute, an organ of the World Health Organization. The city is internationally-known too for its school of advanced nursing and its veterinary school. Industrial research establishments include those of Ugine at Pierre-Bénite, Rhône-Poulenc at Saint-Fons, Progil at Décines and of l'Institut Français du Pétrole at Feyzin (Laferrère, 1970). There are several grandes écoles, the most prestigious acquisition being that of the Ecole Normale Supérieure de Saint-Cloud. Completion of the new library at La Part-Dieu has been followed by the transfer from Paris of the National School for Librarians.

Until 1960 the business heart of Lyon was confined to the restricted tongue of land between the Rhône and the Saône, but following the city's designation as métropole the decision was taken to develop a new commercial and administrative centre in the Brotteaux district to the east of the Rhone. Originally intended for housing, this 28 ha. site of a former cavalry barracks, known as La Part-Dieu, saw its first building completed in 1971. There have since been added some 450,000 sq. m of office space, what claims to be one of the largest shopping centres in Europe with shops on three levels, the Maurice Ravel concert hall with a seating capacity of 2,000, a library which boasts ten reading rooms and a 17-floor book silo with capacity for 2 million volumes ('the biggest literary warehouse in Europe'), main line rail and métro stations, and parking for over 4,000 cars. Many of the financial and business organizations referred to above have their offices at La Part-Dieu which has been described as 'La Petite Défense' and 'Manhattan on the Rhône'.

La Part-Dieu is Lyon's most striking monument to the métropole policy. With the city's métro, it is also testimony to the influence of Louis Pradel, mayor of Lyon from 1957 to his death in 1976. M.Pradel, a supporter of the then-government, was also president of the 56-commune communauté urbaine (la COURLY) and gained a reputation both for his partisan support of the interests of Lyon and his love of concrete. It was he who launched the idea of the métro in 1963, and although work did not begin on it for ten years, the first line, from Villeurbanne to the city centre, was opened in April, 1978 and other sections have followed. Pradel's views did not always accord with those of the more left-wing administrations in industrial suburban communes such as Vénissieux and Villeurbanne, and he was suspicious of the broader interests represented by the OREAM. Strong opposition was expressed to the idea of a New Town and as late as 1975 he was quoted as saying, 'L'Isle-d'Abeau, connais pas' (Le Monde, 1 October, 1975).

It was OREAM's responsibility to plan for the wider urban

region and in 1970 a schéma d'aménagement was approved which
recognized the tendency for Lyon to grow, both in linear
fashion along the Rhône-Saône axis, and also eastwards across
the gravelly plain of Lyonnais in the direction of Geneva and
Grenoble.   The plan was a polynuclear one which proposed four
satellite developments within a radius of approximately 30 km
of the city centre (Figure 5).   L'Isle-d'Abeau and la Plaine
de l'Ain (or Méximieux) were to accommodate populations of
300,000, Belleville-Villefranche and Vienne-Roussillon,
170,000 each.   Both L'Isle-d'Abeau and Méximieux were to be
promoted as New Towns, but in the event it was decided to proceed only with the former which would help to accommodate the
families of the 20,000 or more workers associated with the new
airport of Satolas (Chapter 9).   Development of Plaine de l'Ain
has taken the form of an estate for heavy industry able to take
advantage of nearby facilities such as the Bugey nuclear power
station.

OREAM's zone d'étude also took in Saint-Etienne, but did
not extend to Grenoble, leaving the latter in the ambivalent
position of being a part of the métropole à trois têtes, but
not of OREAM's schéma.   Plans for Saint-Etienne have expressed
three major concerns: (i) problems of employment caused not
only by the contraction of the basic industries of coal and
iron, but also by the difficulties experienced in firms such
as Manufrance, makers of armaments and cycles for the last
hundred years, (ii) the necessity of carrying out urban
renewal in the older industrial quarters, and (iii) the desire
to find space for new housing and employment projects.
Saint-Etienne's confined site has dictated a plan for the
latter involving an extension of the city northwards in the
Plaine du Forez and it is principally here, in the direction
of Clermont-Ferrand, that modern housing and industrial estates
have been laid out.

Grenoble, 'la ville la moins provinciale de province',
is more representative of what Ardagh (1968) calls 'The New
France' than any other French city.   With a population in the
agglomération that has grown from under 100,000 in 1946 to some
383,000 in 1975, it is a cosmopolitan city of scientists,
students and workers in the modern industries of metallurgy
and electronics.

The foundations of postwar growth rest on an established
reputation for hydraulic and electrical engineering on the
part of firms which supply equipment to power stations in the
Alps and elsewhere.   Since the war this base has been widened
as a result of growth, in particular, of industries concerned
with electrical appliances and electronics.   Some, such as
CSF (Compagnie Générale de TSF) have moved from Paris,
encouraged by decentralization policies but also by the
attractiveness of the living environment of Grenoble with its
easy access to the mountains and ski slopes.   Similar reasons

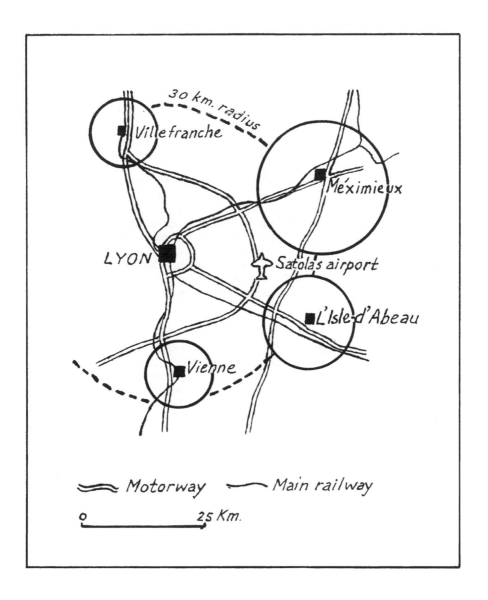

Figure 5.   OREAM plan for greater Lyon

help to explain the rapid expansion of industrial and scientific research, scientists being drawn to the laboratories of the university and to those set up by firms (Neyrpic, Pechiney, Air Liquide) or public bodies such as the CEA (Commissariat à l'Energie Atomique). The centre for nuclear studies was founded in Grenoble as part of a planned decentralization programme, and its arrival in the 1950s has been followed by the setting up of several related establishments. Other organizations contributing to the scientific atmosphere include the Institut National Polytechnique and the Centre National d'Etudes des Télécommunications.

An ambitious programme of public works has had to be carried out in Grenoble in order to accommodate the needs of a population growing at an annual rate of around 4-5%. Provision has also had to be made for the continually expanding tourist industry. The Winter Olympics, held in the city in 1968, brought heavy investment in infrastructure, and this has been followed by much rebuilding and the development of a secondary nucleus, the 'villeneuve' de Grenoble-Echirolles. As usual in French cities, the political factor has not been absent from planning. Since 1965 Grenoble has been a city of the trendy left, more fabian than working class, and there has been a commitment to change and development that is evident in the fabric of the modern city.

Marseille - Aix

Marseille and Lyon evoke very different images in the mind of the average Frenchman. The Lyonnais is regarded as dour and puritanical, interested more in earning money than in outward display. The Marseillais, by contrast, is erratic, exuberant, extrovert; his city is outward-looking and cosmopolitan, associated in the popular mind with immigrants and the Foreign Legion, with exotic crimes and - until 1953 - bad government. Of the two cities, Marseille has been the more 'malade de sa réputation', and it has been one of the aims of planning this métropole to create a better image, more appealing to the investors of Paris and overseas. It is ironical, though perhaps significant, that the mayor of Marseille since 1953, the socialist M.Gaston Defferre, should have been a rather stern protestant from the Cévennes.

Marseille's wealth was founded on the colonial trade and the loss of France's colonial empire over the last 30 years has brought difficult problems of adjustment of both an economic and a social kind. The city has always relied for employment more heavily on commerce than on manufacturing industry and in 1975, for example, had only 62 industrial jobs per thousand population compared with the national average of 110 and figures of 158 in Lyon and 164 in Lille. This narrow base has been weakened further by the contraction of a number of traditional industries, most of them closely associated

with the colonial trade such as soap-manufacture, various food
and clothing industries, and ship repair.   The difficulties,
and the municipal response to them, are well illustrated in the
case of Titan-Coder, a firm which repairs railway equipment and
manufactures trailers and which was threatened with closure in
1974.   The site and buildings were bought by the local
authority and let to a specially-constituted mixed economy
company which successfully took over and continued production.
A similar joint enterprise was set up to rescue the firm of
Griffet, but problems have continued and the threatened
collapse of the ship-repair firm of Terrin became a matter for
national debate in 1978, taking eighteen months to resolve.

The loss of ship repair work was regarded as particularly
serious in view both of the city's long dominance of this
industry and of investment in the port aimed at securing the
future of repair business for Marseille.   In 1971 the
decision had been taken to build a new dry dock capable of
accommodating the largest vessels afloat and 'number 10',
1,500 feet in length, was opened in 1975.   Completion of the
dry dock was part of a programme of works aimed at modernizing
the port following the movement of bulk cargo handling away
from the city towards, first the Etang de Berre, and later the
new port of Fos.   These have included the installation of roll
on-roll off facilities and the creation of new container berths,
the aim being to promote Marseille's advantages as a land
bridge between northern Europe and the Mediterranean world.

In Marseille the attitude towards Fos is as ambivalent as
that of the city fathers of Lyon towards L'Isle-d'Abeau.   The
city and Fos are within the same département, part of a single
OREAM, and served by one port authority, but the creation of
Fos has been viewed with considerable suspicion.   The new port
and industrial area has been seen as representative of outside,
above all Parisian, interests, its development likely to have
a detrimental effect on established businesses in Marseille.
M.Defferre was quoted as saying that 'Marseille ne doit pas
devenir la cité-dortoir de Fos' (Le Monde , 23 November, 1971).
These fears have been somewhat allayed, however, by the failure
of Fos to attract the range of secondary industries that had
been anticipated and there is now a more positive attitude in
Marseille aimed at providing the headquarters offices, services
and other facilities that will be needed when further expansion
eventually takes place at Fos.   Marseille's own manufacturing
sector has also profited from the recent growth of firms making
equipment for underwater exploration and the offshore oil
industry.

Marseille's claim to be the second city of France derives
not from the population of the agglomération, which is less
than that of Lyon, but from the total living within the single
very large commune (23,000 ha.) of Marseille itself.   Little
more than 600,000 after World War II, this had risen to

914,000 by 1975. Foreigners account for around 10% of the total, and <u>pieds noirs</u>, who returned to France after independence was granted to Algeria in 1962, for at least a further 10%. Estimates of the numbers of foreigners vary, depending partly on whether or not they include the children born to immigrant families after their arrival in France who have since acquired 'papers' as French nationals. Inclusion of the whole family would probably raise the 'immigrant' total to around 120,000, at least 80% of them of North African origin or culture. This immigrant community includes some 25,000 célibataires - unmarried workers. The rest are in families, which tend to be large, averaging eight persons.

Provision of cheap housing has been the greatest single task faced by the municipality over the last 20 years. The <u>bidonvilles</u>, shanty towns, of which there were about 30 at one time, have been gradually eliminated - only six remained at the beginning of 1981 - but they have been replaced by what their critics have described as 'bidonvilles verticaux' (Kinsey, 1979). These typically consist of small HLM apartments, often of no more than four rooms, clustered in concrete-built <u>grands ensembles</u>. They, in turn, are grouped in ZUPs, the largest of which is the notorious 'ZUP numéro I' which comprises 10,000 HLM units, accommodating a population of 45,000. There is a heavy concentration of this kind of housing on the cheaper land to the north of the city, accentuating the contrast within Marseille between a proletarian north and a bourgeois, residential south. In some of the <u>grands ensembles</u> the population for which they were originally built in the 1950s has been almost entirely replaced by an immigrant population of North Africans. Social problems are acute, the situation encapsulated in a phrase quoted by Marc Ambroise-Rendu, 'C'est un peu comme le Far West du siècle dernier' (<u>Le Monde</u>, 24 June, 1981).

Faced with the economic and social problems described above, it is scarcely surprising that plans for Marseille should have taken the form of 'un grand programme de petits travaux'. These works have included, in addition to housing, new roads to alleviate traffic congestion, and the provision of social facilities such as schools, hospitals and recreational open spaces, in an attempt to enhance the quality of life in ill-equipped suburbs. Attention has also been given to the improvement of the city's water supply (from the Canal de Provence) and waste disposal system, coupled with the restoration of beaches (Prado) spoilt by earlier methods of removing waste.

There have been two major exceptions to this policy of relatively small-scale improvement. The decision to build a <u>métro</u> was taken in 1969 as part of a new, overall plan for the city and the first section, from the suburb of La Rose to Castellane, was inaugurated towards the end of 1977 (Tuppen,1980).

A scheme of urban renewal in the city centre was begun in 1973, partly in response to what was taking place at Fos, and the first stage of this modernization and rebuilding programme, the Centre-Bourse, was opened in October, 1977. It includes 40,000 sq.m of shopping space, a museum devoted to the city's history, and the preservation in a garden setting of archaeological finds relating to the ancient Phoenician port. Improvement is gradually being extended to neighbouring quartiers - les Carmes,Sainte-Barbe, and the station area of Saint-Charles. As a result of this investment the city has recorded an improved 'image de marque' (Figure 2), but the popular image of the city is still coloured by the violence and racial problems which provide good copy for the national press.

Thirty kilometres from Marseille, the old capital of Provence, Aix, has close functional ties with the city. Growth of employment, particularly in services, helps to explain a doubling of the population from 54,000 in 1954 to 115,000 in 1975. No more than a quarter of the total now live in the 'old town' of Aix-en-Provence, and the modern suburbs share with other fast-growing urban centres the problems of housing, amenities and traffic congestion.

Lille - Roubaix - Tourcoing

Between 1968 and 1975 the population of the commune of Lille fell by 21,000 or 11% to only 170,000. But the city is at the centre of a conurbation that also includes the textile towns of Roubaix (110,000) and Tourcoing (102,000), together with a host of smaller industrial and suburban settlements, and the population of the 87 communes that make up the communauté urbaine totalled 1,115,000 in 1975.

The communauté urbaine was set up in 1968, against the wishes of many of the communes involved, in order to plan for the whole conurbation and avoid the worst consequences of a highly fragmented pattern of local government. It has responsibilities that range from water supply, fire services and waste disposal to the planning of roads, schools and industrial estates. It was 1975, however, before approval was gained for the communauté's first plan des sols, and local rivalries and self-interest have continued to affect the overall direction of the wider urban region. George Sueur, for many years the local correspondent of Le Monde, published a study of Lille-Roubaix-Tourcoing in 1971 which he subtitled, 'métropole en miettes' (literally, metropole in pieces, or fragmented metropolis). He observed, 'Depuis 1963, on accumule les études, les plans, les schémas et on les modifie sans cesse; mais il reste à inaugurer la première réalisation typiquement métropolitaine', (Sueur, 1971).

Problems of housing and transport have added to the difficulties of planning. The metropolis has a legacy of

poor-quality industrial housing.   Small brick-built dwellings,
arranged in rows or courées, are intermingled haphazardly with
factories and stores, giving rise to l'urbanisation anarchique.
Solutions to this housing problem have been made more difficult
by the high rate of natural increase of the population and by
the immigration of foreign workers, leading to heavy reliance
on subsidised dwellings built by organismes d'HLM or comités
interprofessionnels du logement (CIL).   The new town of
Villeneuve-d'Ascq (Lille-Est), work on which began in 1971, has
been viewed with suspicion as an investment likely to attract
public funds to the detriment of urban renewal projects in the
older parts of the conurbation.   Hostility has been strongest
in Roubaix and Tourcoing where there is resentment at what
appears to be a policy favouring Lille.

There is a complicated pattern of daily commuting in the
northern metropolis.   The needs of these travellers were met
in the past by a dense system of tramways, and the 'Mongy'
still provides a link between the centre of Lille and the twin
towns to the north, but otherwise town roads are congested with
the usual mixture of cars and buses.   The fragmented nature
of the conurbation appeared to provide less justification for
investment in a métro than in the case of Lyon or Marseille,
but in 1978 work began on a lighter, and cheaper, version of
métro, the VAL (véhicule automatique léger).   The first line,
part above and part underground, is due to be opened in 1983,
linking the new town with the city centre and the regional
hospital complex to the south.

Lille was the first major provincial centre to be linked
with Paris by motorway (1967), and autoroutes now radiate from
the city to Dunkerque and, via the Belgian network, to many of
the larger towns of north-west Europe.   Postwar investment has
also taken the form of main-line rail electrification and a
deepwater canal link with the coast.   Lille's connection with
the latter was opened in 1976 and a new port created at Santes.
Yet, despite these major schemes, the transport services within
Lille's own regional hinterland remain in need of improvement.
Planning at this regional scale is the responsibility of the
northern OREAM, and a schéma was produced in 1971.   But OREAM's
role is questioned in Lille-Roubaix-Tourcoing where it is felt
that there is favouritism towards the towns of the coalfield
to the south.

Largely as a result of history, Lille has been compelled
to share the functions of regional capital with other towns,
notably Cambrai, seat of the archbishop, and Douai, with its
Cour d'Appel and - until the late nineteenth century - the
state university.   Estienne (1979) observes the relatively
weak role of the local press and the fact that financial
institutions, though numerous, are mostly branches of Parisian
firms.   'La ville dispose de tous les services administratifs
d'une capitale régionale, mais il lui manque un peu de "panache".

48

The 'old town' of Lille is made up of several parts: a Flemish core with typically tall houses, Vauban's seventeenth-century extension to the north, and nineteenth-century suburbs to the south. Gradual desertion by the bourgeoisie had left Lille with a problem of spreading poverty in these historic central quarters, and the search for "panache" and status has taken the form both of restoration and of extensive clearance to create a new commercial centre. Some land was made available as a result of the removal of the vegetable market to the suburb of Lomme in 1972, whilst the most recent phase of this major redevelopment scheme has been a new 'super-station' that will accommodate main-line trains, trams and the métro. The nearby Grand-Place is also undergoing restoration to coincide with the métro's inauguration in 1983.

Investment in Lille's centre directionnel has added to a sense of neglect in Roubaix and Tourcoing where both towns have suffered from the severe contraction that has taken place in the wool textile industry. Sueur, in 1971, drew attention to a widening division within the métropole: ' ... d'un côté, la capitale lilloise modernisée et embellie, flanquée de sa ville nouvelle: Lille-Est; de l'autre, le faubourg industriel' (of Roubaix-Tourcoing). With problems of unemployment, bad housing and a large immigrant community, the twin towns have gained little from the new industrial developments that have tended to favour locations on, or close to, the coalfield. Their response, an unusual one for local authorities, has been to set up (in 1979) a société d'économie mixte which draws on both public and private capital in order to equip sites for incoming firms. The Chamber of Commerce has also financed the prestigious Mercure office tower in Roubaix, where renewal of the Alma-Gare quarter may be seen as an attempt to offset the growing commercial dominance of Lille.

Bordeaux

Bordeaux, like Lyon, used to be regarded as a solid, bourgeois, provincial city, its 'Chartrons' wine empires having a similar reputation to the old Lyon families of the 'soyeux'. As in Lyon, much has happened to change that image since the early 1960s.

Between 1962 and 1975 the population of the agglomération of greater Bordeaux grew by more than 100,000 from 507,000 to 612,000. During the same period, that of the central ville of Bordeaux itself fell by some 50,000 to 224,000. The fact that the central city administration was responsible for 50% of the agglomération's population in 1962 but only 36% in 1975, highlights the need for cooperation in planning matters between the city and its surrounding communes during a period of growth. One of the ways of ensuring such cooperation is the strategy of the communauté urbaine, and this has been utilized in Bordeaux as it has in Lyon, Lille and Strasbourg. Here, 27 communes were united for planning purposes at the beginning of 1968.

In all four cities the communauté urbaine was a solution imposed with the backing of central government in order to direct the expansion of the metropolis. Success thus depends to a large extent on the personalities involved, and the cooperation achieved in greater Bordeaux, even before the communauté urbaine was set up, owes much to the skill in negotiation of the city's long-serving mayor, Jacques Chaban-Delmas. M.Chaban-Delmas was also, successively, a government minister, President of the National Assembly and Prime Minister during the growth years of the 1960s and early 1970s, and the city gained enormously from his contacts in national and international circles and his ability to represent local interests at the highest level. This is best illustrated in the case of the Ford gearbox factories, opened in 1973 and 1976 and now employing 4,000 workers, which were built on the outskirts of Bordeaux in competition with many other French cities. Other examples may be found in the aircraft and aerospace industries, and in firms manufacturing chemicals and pharmaceuticals. Since 1960 the agricultural and colonial base of Bordeaux's industry has been transformed by the influx of these modern activities. The link between decision-making and urban growth is well illustrated in the following comment of Dumas (1978):

> Durant la période qui va de 1962 a 1974, le maire de Bordeaux a établi une suprématie incontestée sur l'agglomération, grace à une remarquable efficacité économique. C'est par lui qui passent les grandes négociations qui amènent le développement de l'aéro-spatiale, aussi bien que celui de la construction auto-mobile ou électronique. C'est lui qui est le médiateur entre les milieux économiques locaux et les instances nationales de l'aménagement du territoire ... C'est lui enfin qui met en place une fédération des communes de banlieue.

Most of the new factories are in the outer suburbs, where industrial estates have been built on former forest or marsh-land beyond the ring motorway (Figure 6). There is a concentration to the north of the city at Blanquefort, Bassens and Ambès and, at the north of the estuary, Le Verdon. The oil port of Le Verdon was opened in 1967 and the nearby Pauillac refinery four years later. It was intended that oil-based chemical industries sited here would compensate for the eventual decline of the Lacq gas fields. That hope has not been realized but there has been progress since 1976 with a shift to general merchandise and container traffic.

It is not only manufacturing that has moved to the suburbs. The decision was taken in the mid-1950s to move the university

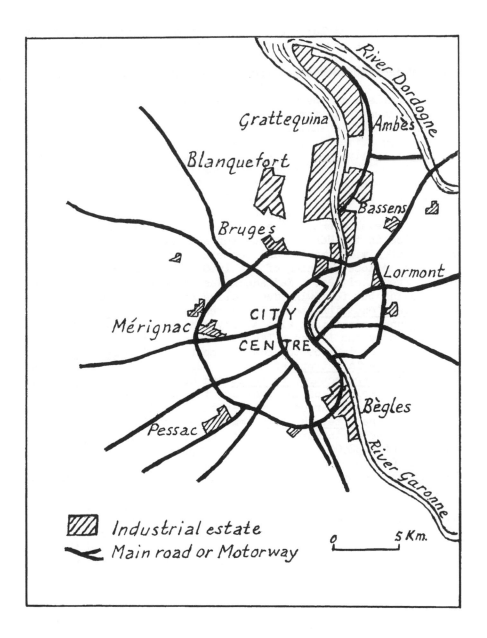

Figure 6. Bordeaux's industrial estates

to a new campus at Talence in the south of the city. With its 280 ha. this is now one of the largest in France and other educational establishments have clustered in its vicinity. Another decision concerned the International Fair, formerly held in the centre of the city on the Quinconces Square. Some sandy swampland to the north of the city was acquired in 1959 as a permanent site for the Fair and work followed on the creation of the 160-ha. lake which has given the locality its name - le quartier du lac. A mixed economy company, representing the city, the Fair authority and local banks, was set up to direct the development, and most of the exhibition buildings were completed by the late 1960s. Hotels, offices and a range of recreational facilities have followed.

The most ambitious project, however, has been one of central redevelopment. The clearance of 26 ha. of decayed properties in the Mériadeck quarter and their replacement by a complex of administrative and commercial buildings is typical of the grandiose planning of the 1960s as it affected the larger French cities. Mériadeck was to have been redeveloped for housing but following designation of Bordeaux as a métropole d'équilibre the plan was changed to a centre directionnel consistent with a city that was expected to grow to 800,000 by the 1980s. The government contributed 100 million francs in 1969 towards the cost of this quartier d'affaires de la métropole, the new buildings of which include the regional préfecture, rectorat and offices of the communauté urbaine, as well as of the post and telecommunications service. These are arranged around a tree-planted mall which extends the grounds of the hotel de ville immediately to the east. Completion of the administrative units has been followed by the building of a 32,000 sq.m shopping complex.

The growth of Bordeaux slowed in the late 1970s, partly in response to political changes that have weakened the pull of Bordeaux, both nationally and also locally in the city's dealings with increasingly leftward-leaning suburbs, and partly in response to the swing in public opinion away from big schemes like that of Mériadeck. The rebuilding and gentrification of this latter district is in contrast with the resistance that has been put up more recently to urban renewal in the nearby working class and immigrant quarters of Saint-Pierre and Saint-Michel. The legacy of the last quarter of a century is nevertheless clear to see and for his contribution to this, the mayor of Bordeaux has been compared with Tourny, the Intendant who remodelled and beautified the city in the eighteenth century.

Toulouse

In the course of a one-day visit to Toulouse in May, 1971, during which he flew in a prototype of Concorde, President Pompidou declared that Toulouse 'est un des lieux privilégiés de la France moderne'. It was a remark intended to please those who, in the 1960s, had sought to emulate 'l'expérience de Grenoble' and make Toulouse one of the leading centres of advanced engineering and scientific research in France.

A measure of the city's progress is to be found in the growth of population which, from a total of little more than a quarter of a million in the early 1950s, had risen to 370,000 in 1962, to 447,000 in 1968 and 506,000 by 1975. The totals refer to the 36 communes making up the agglomération of greater Toulouse, but the ville itself is broadly defined (12,000 ha.) and accounted for 374,000 of the half-million recorded in 1975. Growth was particularly rapid in the 1960s when more than 10,000 people a year were being added to the city's population, much of this as a result of migration. Repatriates from North Africa alone added about 25,000, accounting for around one-fifth of the total growth.

The economic foundations of what has happened in Toulouse are to be found in the aircraft and chemical industries established there in the early years of the present century, especially during and soon after the First World War. Many thousands are now employed by SNIAS (Aérospatiale), Dassault-Bréguet and other companies in the building, repair and testing of aircraft for both civilian and military markets. Extensive factories and testing grounds are to be found to the north-west of the city, near Blagnac airport, accounting for the transformation of nearby suburbs such as Colomiers. But the impact of the politique des métropoles d'équilibre arises less obviously from the expansion of these established manufacturers than from the growth of the associated research side of aircraft engineering and the widening of the scientific base to take account of progress in the closely linked industries of electronics, satellite development and telecommunications. This has been encouraged by the government, notably by the transfer of institutes of higher education and research from Paris, but also through the encouragement given to foreign companies to establish factories in Toulouse. The latter include branches of several American electronics firms, for example, Motorola, makers of semi-conductors, who settled at Le Mirail in 1967.

The city's ambition to be 'la capitale française (even "européenne") de l'aéronautique et de l'espace' is symbolized in the scientific complex of Rangueil-Lespinet which straddles the Canal du Midi to the south-east of the city. Begun in 1963, this campus has acquired a range of laboratories and advanced teaching establishments, many of them grouped around Paul Sabatier University, once the Faculty of Sciences of the University of Toulouse. The key to the decentralization programme was the move to Toulouse from Paris in the 1960s of several prestigious schools of aeronautical engineering, together with their closely related research establishments. They include the Ecole

Nationale Supérieure de l'Aéronautique et de l'Espace
("SUP'AERO"), the oldest of its kind in France, the Ecole
Nationale de l'Aviation Civile, and the Ecole Nationale
d'Ingénieurs de Constructions Aéronautiques.   In addition to
these schools' own research laboratories, there is the Centre
Spatial de Toulouse, opened in 1968 as a branch of the Centre
National d'Etudes Spatiales, itself set up in 1962 in order to
promote the French space research programme.   Widening the
range of scientific enquiry, the city's list of higher education
establishments also includes schools of chemical and electronic
engineering and of veterinary science.

More recently, a number of small government departments
have been encouraged to move to Toulouse: la Direction Tech-
nique des Télécommunications, le Service Statistique du
Commerce Extérieur, and la Météorologie Nationale.   The
latter's move is logical in terms of the connection between
space research and the use of satellites in meteorology.

Rivalling Rangueil-Lespinet as symbol of the new Toulouse
is the satellite township of Le Mirail, built within the city
boundary some four km south-west of the centre.   Conceived
in the late 1950s, Le Mirail is being developed by a mixed
economy company as a zone à urbaniser en priorité (ZUP).   It
differs from the usual ZUP, however, in its scale - 800 ha.
and 23,000 dwellings, planned for a population of 100,000 -
and in the attention given to layout and design by its
architect, Georges Candilis.   The Y-shaped blocks of apartments, arranged in what is described as a varied linear structure, represent a departure from the monotonous blocks that
characterized the early grands ensembles.   The plan also
emphasizes pedestrian movement, with a raised walkway running
through the centre of the township and 'streets in the air'
at fifth, ninth and twelfth storey levels to avoid the sense
of isolation often felt by tower-block dwellers (Figure 7).

Work began on the site in 1960 but was slow initially,
speeding up after Toulouse was designated a métropole
d'équilibre and in response to the rapidly growing population
of the city.   The first houses were occupied in 1966 and by
the early 1970s the southern portion of Le Mirail, known as
Bellefontaine, was substantially complete.   Motorola and the
computer firm of CII (Compagnie Internationale pour
l'Informatique) were established on a nearby industrial estate.
Since that time, however, progress has slowed and there have
been criticisms of the high proportion of 'social' (HLM)
housing and of the poor integration of shopping facilities
with the housing.   More attention is now being paid to low-
rise and private housing as development spreads towards the
focal point in the northern part of the township, the campus
of the University's Faculty of Letters.   Looking back, Lévy
(1977) sees Le Mirail as a typical product of the expansionist
philosophy of the 1960s: 'Le project de Toulouse-Le Mirail

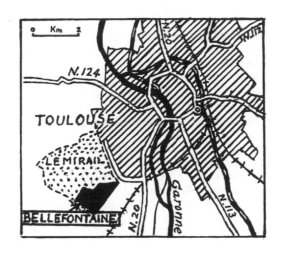

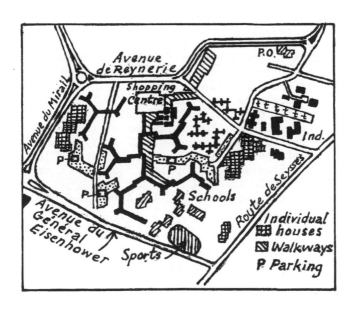

Figure 7. Le Mirail and Bellefontaine quarter

55

illustre à la fois une conception de l'architecture de masse
et une philosophie de l'urbanisme dans la conjecture historique
des années 1960, où prospérité, croissance, et progrès
semblaient étroitement associés'.

Equally typical of its time was the treatment accorded to
the quartier Saint-Georges, a small working-class district of
the 'old town' that was largely cleared in the 1960s.   The
intricate pattern of narrow streets has here given way to two
open squares around which are arranged offices and expensive
apartments.   Public opinion has since moved away from this
kind of clean-sweep planning, however, and it has become more
common to see urban renewal attempted without displacing the
resident population, for example in the quartier Saint-Aubin.
Yet pressures remain strong for change at the centre, evident
in plans for the Compans-Capparelli barracks site where it is
proposed to build municipal offices and a cultural and sporting
complex.

Toulouse exemplifies well the conflicts and problems that
accompany rapid growth.   The population of the historic core
fell from 51,000 in 1946 to only 30,000 in 1975, making adjustment of some kind inevitable.   Over the same period the brick-
built quarters of 'la ville rose' have become engulfed in a
widening band of concrete apartments, prefabricated factories
and off-the-peg pavillons.   An increasing proportion of this
growth now takes place outside the boundaries of the ville,
and with no district or communauté urbaine to direct development, planning has been slow and difficult.   Boasting the
highest rate of car ownership in France, with an average of
more than one vehicle to each household, Toulouse also suffers
from severe traffic circulation problems.   Finally, observe
that Toulouse now accounts for a quarter of the entire population of its region, Midi-Pyrénées (12% in 1936) and is ten
times the size of the next largest town, Tarbes.   'La ville
rose, qui devait être une "métropole d'équilibre", est en train
de devenir à l'échelle du Sud-Ouest, une tête hypertrophiée sur
un corps de plus en plus débile' (Marc Ambroise-Rendu in
Le Monde, 4 March, 1981).

Nantes - Saint-Nazaire

The métropole of the lower Loire extends downstream for
some 60km from Nantes, the dominant partner, with a population
of half-a-million (260,000 in the ville) to Saint-Nazaire
(60,000), a town founded only in the middle of the nineteenth
century and dominated ever since by the fortunes of the shipbuilding industry.   The necessity of coordinating planning
for the whole estuarine sub-region was recognized in the setting
up of an OREAM in 1966 and the schéma directeur for this aire
métropolitaine of Nantes - Saint-Nazaire was published in 1970.
Its main objectives were:
    (i)   creation of a new port and industrial zone centered

on Donges-Montoir,
(ii)   improvement of transport links, including a bridge
        across the lower estuary,
(iii)   promotion of high order service activities at Nantes.
    Saint-Nazaire's shipbuilding industry, largely concentrated
on the yards of the Chantiers de l'Atlantique, benefited from
the oil boom of the 1960s.   A new dock, capable of building
the largest tankers, was completed in 1968, and several vessels
of half-a-million tonnes have been built here, together with
methane carriers.   Orders have fallen off since 1974, however,
and in 1976 the Chantiers combined with Alsthom to form the
Alsthom-Atlantique company, widening its activities to include
other branches of the engineering industry.   Employment in the
shipyards fell from over 10,000 in 1956 to just over 6,000 in
1979, but another 3,000 were employed in Alsthom-Atlantique's
engineering division now centered on the new industrial estate
at Montoir.   The oil industry has also grown with the completion of the port and refinery at Donges and the new (1981)
terminal for imported methane at Montoir.   The bridge over
the estuary was opened in 1975.
    The survival of shipbuilding in Nantes is now dependent
upon the one company of Dubigeon-Normandie, builders of
specialized vessels including car ferries, and numbers employed
in the city's shipbuilding industry fell from 7,000 to only
2,000 between 1956 and 1979 (Cabanne, 1979).   But the overall
employment base is wider than that of Saint-Nazaire and
includes firms set up as a result of decentralization policies
(electrical engineering, tyres for Renault).   Nantes also
has a large tertiary sector, swelled since 1966 by the transfer from Paris of personnel employed by the Foreign Ministry,
and by the establishment here of the Institut Scientifique et
Technique des Pêches Maritimes, and an agence régionale of the
IGN (Institut Géographique National).   The physical expansion
of the city beyond the boundaries of the ville has brought
problems similar to those experienced in other métropoles.
There is a Syndicat Intercommunal Multiple de l'Agglomération
Nantaise (SIMAN) which brings together the communes to discuss
matters of common interest, but no formal communauté urbaine
has been set up.   Problems of traffic circulation include those
of crossing the river and the respective merits of a new bridge
or tunnel have been discussed exhaustively.   To ease commuter
movement a new tramway, built by Alsthom, is to open in 1983,
making Nantes the first French city to re-adopt this system.

Nancy - Metz - Thionville
    The Lorraine métropole is unique in having no single,
dominant urban centre.   The aire métropolitaine (an OREAM was
established for planning purposes in 1966) has a population
approaching three-quarters of a million but this is divided
between numerous small towns as well as the three main urban

foci.   In 1968, when the OREAM produced its 'White Paper' on
the metropolis, the population of the three agglomérations
was: Nancy 258,000 (ville 123,000), Metz 166,000 (ville
108,000), Thionville 136,000 (ville 37,000).   By 1975 the
population living in the rather more youthful commune (or
ville) of Metz had actually overtaken the total in the closely-
defined Nancy (108,000 in 1975; only 96,000 in 1982).

Nancy and Metz, though only 60 km apart, are in separate
départements and have had a very different history which has
created a deep-seated rivalry, summed up in the phrase, 'soeurs
ennemies'.   Neither town has a strong manufacturing sector,
employment depending to a large extent on their service
functions.   Higher education, banking and hospital services
are prominent in Nancy; wholesaling, administration and services
for the military are more evident in Metz.   Thionville, by
contrast, is at the heart of the iron and steel-making district
of Lorraine and is heavily dependent upon this activity.

The valley of the Moselle provides an obvious bond between
the separate towns, and plans for the aire métropolitaine have
stressed the need for good communications in order to take
advantage of this physical link.   The métrolor, a regular fast
rail service between Thionville, Metz and Nancy, was introduced
in 1970.   The motorway joining the three towns was opened in
1972, and the Moselle has gradually been canalized to
'European' standards as far south as Neuves-Maisons.   In a
region that has suffered heavily from the decline of its basic
industries, it is also the valley of the Moselle that has
offered the most attractive sites to new industries lured by
various grants and tax concessions.   These include Kléber-
Colombes at Toul, Citroën at Metz, Renault and a number of
smaller factories in the neighbourhood of Thionville.

In spite of investment aimed at uniting the towns of the
Moselle valley, there have been times when the whole concept
of the métropole has been undermined by jealousy and suspicion.
The population of Nancy felt slighted when Metz was chosen as
seat of the regional préfecture in 1969, and the OREAM felt it
necessary to establish its offices in the neutral town of
Pont-à-Mousson, mid-way between the rival centres.   It also
set up separate planning commissions for the northern and
southern portions of the métropole.   But the route of the
motorway from Paris to Strasbourg proved to be the most conten-
tious issue, both Nancy and Metz seeking to gain the advantage
of an autoroute passing close to the town.   Alternative
strategies were also put forward, for a line that would run
mid-way between the towns, and for a motorway with two branches
in order that both might be served by it.   When the route to
the north of Metz was chosen in 1970, the government was
accused of ignoring the métropole in their anxiety to promote
links between northern Lorraine and the Saar.   Interests in
Nancy, where M.Servan-Schreiber was elected Radical député in

1970, were only soothed by promises of an accelerated programme
of improvement to the R.N.4 and of the motorway along the
Moselle valley.

Local disenchantment with the métropole idea is evident
from a comment in the OREAM's schéma directeur, published in
1970: 'Si les villes qui composent la région évoluent de
manière indépendante, commes elles le firent dans le passé, la
Lorraine aura quelques agglomérations urbaines de taille
respectable, elle n'aura pas de métropole'. Two geographers
expressed the matter succinctly: 'Taille modeste, agglomérations
dispersées et rivales, dynamisme démographique ralenti,
fonctions mal équilibrées, urbanisme d'insuffisante qualité,
mentalités différentes, concurrence des villes étrangères
voisines, les aptitudes de Nancy-Metz-Thionville à former une
métropole d'équilibre paraissent limitées' (Thouvenot and
Wittmann, 1970). To a local politician, 'la métropole
lorraine n'est qu'un leurre' (lure) (J.Feidt, quoted in Le
Monde, 2 July, 1975).

Efforts to achieve greater unity continued in the 1970s,
mainly on the part of OREAM which, in 1973, was responsible for
the creation of an établissement public foncier for the
métropole, the first of its kind outside Paris and the Seine
valley. Its members are drawn from local government, industry,
chambers of commerce, etc., and it is able to levy a special
tax, to acquire land compulsorily, and to make loans in order
to carry out the intentions of OREAM's schéma d'aménagement.
The Paris-Strasbourg motorway was completed in 1976, leaving
as a major unresolved issue, only the site of an enlarged airport to serve the whole of the métropole.

Strasbourg

Strasbourg has the smallest population total of the eight
métropoles. Little more than 300,000 when designated in 1964,
this had grown to 363,000 by 1975 (257,000 in the ville).
Like the other cities, Strasbourg has its housing estates on
the periphery, a major new hospital (Centre de Hautepierre),
and a central redevelopment scheme with offices, shops and
hotels (Place des Halles) but, without the problems caused by
a large immigrant population or a legacy of poor-quality housing,
the pace of change has been slower.

To the traditional industries of food and drink, clothing
and leather, there were added vehicle plants and oil refineries
in the 1960s. General Motors opened its factory for the manufacture of gearboxes and automatic transmissions in 1968, and
growth of the refinery and associated chemical industries
followed completion of the oil pipeline from the Mediterranean
coast in 1963. The port (13 million tonnes in 1977) also
profited from completion of the Grand Canal d'Alsace in 1967.

But the uniqueness of Strasbourg lies principally in its
position on the Rhine, at the heart of the European Community,

and this is evident in the city's quest for a European role, both political and cultural. Since the Council of Europe was established in Strasbourg in 1949, the city has become the seat of other institutions, including the European Court of Human Rights, and the meeting place – not yet permanent – of the Parliamentary Assembly of the EEC. The annual fair is long-established, but the Maison du Commerce International was the first of its kind to be opened in France. A Palais des Congrès was completed in 1973 to serve as a centre for major conferences and to enhance the city's reputation as a centre for music and concerts.

The Métropoles in Retrospect

Manuel Castells (1978) considers 1963 to have marked a turning-point in the evolution of postwar urban policy in France. The years from 1951 to 1963 saw the emergence of an 'urban social policy' characterized by the development of a large subsidized housing sector (HLMs) and by the introduction of public controls to regulate the way in which land was used and to avoid land speculation (ZACs, ZADs). By the early 1960s, however, there was growing disenchantment with a social policy that had produced the grands ensembles, and the years between 1963 and 1973 brought a reaction in the form of 'preferential aid to monopoly capitalism'. The Schéma Directeur for Paris (1965) was a product of these years as was the politique des métropoles d'équilibre.

In support of his argument, Castells cites the investment that took place in urban renewal projects, e.g. La Part-Dieu, interpreting this as a response to the need for more offices and services in order to assist the management of capital during a phase of sustained economic growth. Public funds were also used to aid capital accumulation through investment in the urban motorways and métros which enable the workforce to gain access to these new directional centres. Ringroads, likewise, have proved an attraction to the stores and hyper-markets of the major commercial organizations.

Other commentators, less Marxist in their approach than Castells, have also criticized the scale of development, referring to the 'megalomania' of the 1960s which put the emphasis on prestigious schemes rather than seeking to satisfy the needs of the local community. Priority was given to accommodating the multi-national firm, the grande surface commerciale and the super leisure centre, rather than encouraging craft industries, aiding the survival of local shops or making recreational provision for the housing estates. Old houses were too readily described as vétustes and taudis, to be cleared in order to make way for smart office blocks or apartments. Only in the 1970s, as people have come to regret the loss of the old vie du quartier, as well as the destruction of historic buildings, has there been

60

a reaction and change of policy towards conservation.

France was not unique, of course, in the kind of emphasis placed on urban development in the 1960s. 1963 saw the publication of Buchanan's Traffic in Towns, and city planning in many countries sought to satisfy the needs of the motor car, the office block and the shopping centre. But the hegemony of Paris produced a unique response in the form of the French politique de métropolisation which, by the emphasis which it placed on building up selected cities as countermagnets to the capital, certainly did nothing to discourage the large-scale, monumental and prestige-seeking approach to development.

Criticisms of gigantisme aside, it remains to be considered whether the métropoles have achieved their primary objective of countering the dominance of Paris. Certain tentative conclusions may be presented, bearing in mind the comment of Pumain and Saint-Julien (1978) that 'jusqu'à présent, aucun bilan d'ensemble de cette politique n'a été proposé' (p.97).

It is the presence of high-order services, particularly those involving decision-making, that one must seek in order to judge the success of the politique. Progress has, in this respect, been limited however. The métropoles can boast considerable advance in the provision of higher education and research facilities, and in medical and shopping services, but there has been no strong movement of headquarter offices to the provinces and, despite the presence of some foreign banks and easier loan facilities, the financial independence of these cities is still constrained. Barrère and Cassou-Mounat (1980) comment on the provision of 'tertiary' facilities: 'Au total, les métropoles régionales ont acquis dans le domaine des grands équipements un poids plus important, sans cependant équilibrer, en volume et en niveau de spécialisation le pôle parisien'.

The métropoles have gained considerably from investment in transport infrastructure, but to an unequal extent. In the early 1980s there were still gaps in the motorway network between Nantes and Paris, Bordeaux and Toulouse, Nancy and Lyon, and by no means all the cities are served by an airport of international status. In manufacturing, some of the cities, notably Bordeaux and Toulouse, have been successful in acquiring modern high-technology industries, and foreign firms have set up branch plants, but these trends have tended to reduce the level of local control and direction. Pumain and Saint-Julien (1978) observe that the employment structure of the eight métropoles has tended to converge; Lyon and Lille, which were the most industrial, have acquired a higher proportion of 'tertiary' occupations, whilst Bordeaux, Toulouse Nancy and Strasbourg have progressed industrially. They conclude that 'convergence vers un modèle "grande ville" est particulièrement sensible pour les métropoles d'équilibre', but are inclined to believe that such convergence is a consequence more of the initial size of these places than of any set of policies for them.

Since the <u>politique des métropoles</u> was introduced, considerable efforts have been made on the part of the municipalities involved to promote cultural activities. Concert halls and theatres, museums and festivals bear witness to the success of this policy, but expense does not cease with completion of the buildings, and the finance needed to maintain a level of cultural provision conceived in the 1960s is proving a burden to the <u>métropoles</u> in the straightened circumstances of the 1980s.

Another legacy of the past is that of city plans, often drawn up when population projections seemed to justify ambitious schemes of development and renewal. According to Barrère and Cassou-Mounat, 'Le problème essentiel des métropoles régionales semble être aujourd'hui le ralentissement de leur croissance, qui remet en cause certains des grands programmes d'urbanisme élaborés il y a une dizaine d'années'. A decade ago it was still supposed that Bordeaux, for example, would reach a population of 750,000 by 1985, and that three-and-a-half million people would be living in the <u>aire métropolitaine</u> of Marseille by the end of the century. The fall in the birth rate and the halt to overseas migration have made such totals increasingly unrealistic, yet developments have taken place, and are still planned, on the basis of such figures.

A final comment must concern the <u>métropoles</u>' ability to absorb investment without detriment to the remainder of the region of which they are a part. A publication of the OREAM de Lyon sets out as a primary aim of the <u>métropoles</u> that of improving 'leur pouvoir d'attraction sur les régions voisines, non pas pour les vider mais pour recueillir les populations qui émigreraient en tout état de cause et risqueraient de se diriger vers la région parisienne'. That may be considered an honest statement in the light of the development that has taken place in the <u>métropoles</u> but, by putting the emphasis on growth at the centre, it tends also to support those, like Castells, who believe that investment of the kind proposed in the 1960s was bound to add to regional imbalance. There were others, however, who thought that growth of the <u>métropoles</u> was a necessary forerunner of the diffusion of investment throughout the region, beneficial 'spread' effects exceeding negative 'backwash'. To see whether these views were justified, we must next turn to the small and medium-sized towns of the urban system.

Chapter 4.

PLANNING THE URBAN HIERARCHY (II)
THE MEDIUM-SIZED AND SMALL TOWNS

Between 1973 and 1979, the councils of seventy-three
medium-sized French towns entered into a contract with the
government under the terms of which the latter would help to
finance an agreed programme of urban improvements (Figure 8).
This politique des villes moyennes which, for a few years,
short-circuited the existing planning system, can be seen as a
reaction to the gigantisme of the 1960s with its emphasis on
large-scale developments, including those planned for the
métropoles d'équilibre.   It also reflects a growing interest
in 'quality of life' issues, including a concern for the
architectural inheritance of these representatives of a more
traditional France.   Since 1979 the contrat d'aménagement has
no longer been used as an instrument for promoting the ville
moyenne, but programmes of improvement are still undertaken
in the medium-sized towns which are able to draw on the finances
of the Fonds d'Aménagement Urbain, set up for this purpose in
1976.

Concern for the Medium-sized Town
The need for a policy in favour of the medium-sized town
was widely canvassed in the early 1970s.   The Sixth National
Plan (1971-75), for example, referred to the need to make
better use of the development potential of small and medium-
sized towns.   At the same time, the advantages of these
places was being promoted by representatives of DATAR and the
latter's research organization, SESAME (Système d'Etudes pour
un Schéma d'Aménagement de la France).   Addressing a congress
in Bordeaux in October, 1971, M.Jérôme Monod, head of DATAR,
noted a greater willingness on the part of the larger industrial undertakings to divide their organizations into smaller
units of production of perhaps a thousand employees each,
which might be located in medium-sized towns.   For the firm,
modern communications technology ensured that branch plants
remained in easy contact with headquarters, whilst the town
gained a valuable new source of employment from the decentra-

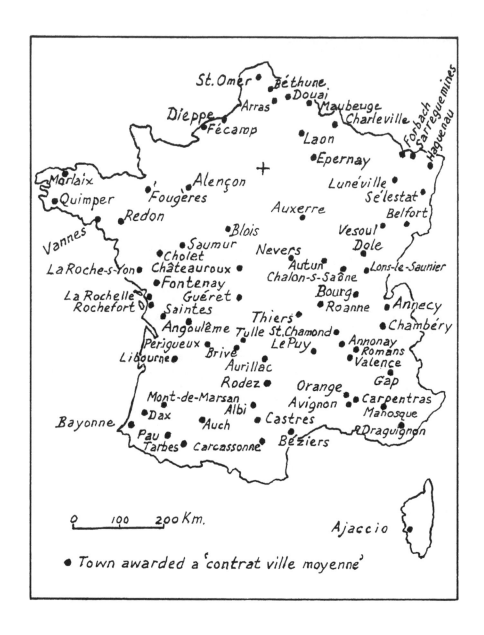

Figure 8. Towns entering into a 'contrat ville moyenne',
1973-79

64

lized offshoot. He also drew attention to the social benefits of the medium-sized towns where new housing could be built without recourse to the tower blocks that had attracted such criticism in the larger cities.

Much emphasis was placed on the quality of life afforded by the medium-sized towns: their picturesque sites and historic buildings, open space and access to the countryside, freedom from noise and traffic congestion, a sense of community in place of the growing anonymity experienced in the major cities. In a speech, reported in Le Monde (14 April, 1972), M.Albin Chalandon of the Ministry of Equipment commented, 'Elles sont souvent belles, toujours riche en complexité et de singularité. En étroite relation avec leur petite région, elles sont réellement des cités'. The social advantages deriving from the retention of a sense of community were stressed by the Prime Minister, M.Messmer, in his closing speech to the Congrès National des Economies Régionales et de la Productivité (CNERP) held at Nice in October, 1972, the theme of which had been 'les villes moyennes, animation et développement'. 'Je vois, enfin, dans les villes moyennes, le terrain d'élection du renouveau de notre vie locale ... Dans une collectivité humaine limitée, les hommes connaissent non seulement leurs voisins et leurs partenaires de travail, mais aussi ceux que le hasard de la vie quotidienne leur fait côtoyer. Ils peuvent choisir en meilleure connaissance ceux d'entre eux qu'ils chargent de gérer la cité' (Le Monde, 24 October, 1972).

Another argument advanced in favour of the medium-sized towns was that they did not suffer from the diseconomies of size that were becoming apparent in the bigger centres of population such as Lyon and Marseille with their need for métros and other major schemes of infrastructural improvement. The costs incurred in cities of different sizes formed an important theme in the report by Joseph Lajugie entitled Les Villes Moyennes which was presented to the Conseil Econo-mique et Social in 1973 (and subsequently published by Cujas under the same title in 1974). Lajugie drew on a variety of sources, including the enquiry into le coût des infrastructures urbaines en France which had been carried out by CERAU (Centre d'Etudes et de Recherches sur l'Aménagement Urbain) for the Commissariat au Plan and published in 1970. He found diseconomies of scale to be most evident with regard to urban transport - the cost of roads, parking and public transport - and least in connection with certain public utility (especially water supply) and educational services. Prud'homme (1973) comes to the same conclusion. Under the heading of social costs, Lajugie noted the generally higher expenses incurred in the larger cities with regard to land and housing, travel, accidents and pollution. Overall there was a clear advantage in favour of the medium-sized town, and he compared his findings with those of 'anglo-saxon' authors whose work tended to reveal the lowest cost of urban services in the size range 20,000-250,000 (e.g. Alonso, 1971).

In planning circles a policy for the medium-sized towns was seen as a logical extension of that for the métropoles d'équilibre, actively promoted in the 1960s. If, as M.Monod had suggested, the larger businesses were willing to redistribute a part of their organization in favour of the villes moyennes, then the latter might act as centres relais, links between the major urban centres and the small towns and villages of rural France. But they would have to be helped to play this role. Otherwise there was a danger that the concentration of investment that had taken place in the métropoles would only reproduce at the regional scale the problem that Gravier had described in Paris et le Désert Francais.

This latter point was made by M.Olivier Guichard, Minister responsible for planning, when he opened the offices of an insurance company in Lyon in October, 1972: 'Il ne faut pas que l'on retrouve dans les grandes villes de province les excès de centralisation que l'on dénonce dans la capitale. Lyon, par exemple, ne doit pas faire preuve vis-à-vis de "sa" région du même "impérialisme" qui caractérise l'attitude de Paris vis-à-vis de la France' (Le Monde, 23 October, 1972). In some parts of France the imbalance was already considerable. Midi-Pyrénées, for example, had the rapidly growing city of Toulouse but no other towns with a population that exceeded 50,000. The alternative to this imperialism of the regional capital was seen to be a well-balanced hierarchy of urban centres and the theme of 'une armature urbaine équilibrée' was a popular one in the development programmes published in the early 1970s by the regional CODERs (commissions de développement économique régional). It was consistent, too, with the legislation of July 1972 which sought to give greater authority in planning matters to the regions.

To the planner the concept of the centre relais embraced not only the idea of downward diffusion of employment from the regional capital, but also upward migratory movement of young adults from the village to the city. The latter was frequently stepped migration, the medium-sized town acting as a kind of staging-post in a chain of moves that had traditionally ended for many in Paris. We have already observed the rapid growth of the medium-sized towns in the 1950s and 1960s (Tables 5 and 6, Chapter 1). It could easily be demonstrated that a large proportion of this growth was at the expense of rural France. A study for INSEE (Institut National de la Statistique et des Etudes Economiques) of the intercensal period 1954-62 (Schiray and Elie, 1970) looked at towns with a population of between 20,000 and 100,000 and this showed an overall gain of 290,000 made up of a net loss of approximately

100,000 persons to larger urban centres, but a net gain of
86,000 from smaller towns and of 304,000 from rural communes.
The latter figure is little less than the number of rural
migrants moving direct to cities of more than 100,000 population (395,000). The totals were seen to support the reality
of stepped migration, whilst the ratio of 4:1, migratory gain
to loss, on the part of the towns of 20,000-100,000 argued a
case for policies to try and arrest the movement of young
people at this stage in the migratory chain.

It was argued that, without policies to improve both
employment prospects and quality of services, the population
would simply drift to the larger cities. In their analysis
of the French urban system, Carrière and Pinchemel (1963) had
singled out the towns of 50,000-100,000 inhabitants as
displaying 'signes pathologiques', stressing their functional
weakness and ill-defined role in the urban hierarchy (Chapter
2). Viewed from Paris, their image was not a favourable one.
Lajugie (1974) refers to traditional attitudes which regarded
the ville moyenne as symbol of mediocrity and 'le trou de
province, où l'on vit dans l'obscurité et dans l'obscurantisme'.
Speaking of the virtues of the medium-sized towns, even
M.Chalandon felt obliged to describe them collectively as 'les
belles endormies' - sleeping beauties.

The need to diversify the employment base of the medium-
sized towns was regarded as an essential element of any policy
towards them. Manufacturing industry was often specialized,
dominated by a particular trade and sometimes by a single firm.
In many of the towns, however, there was a heavy dependence,
not on manufacturing, but on the service sector and for the
smaller ones this meant reliance upon public services, the
professions being strongly represented only in the larger
centres (over 100,000 population). The extent to which some
towns had come to depend on growth of the public service sector
is brought out well in a number of studies of villes moyennes
in south-west France carried out under the direction of
Professor Bernard Kayser of the University of Toulouse-Le
Mirail. The same studies illustrate the complexity of
migratory movement, with foreign workers and French families
returning from Algeria and elsewhere adding to the influx of
rural migrants and of government employees. Rates of population growth were high as a result of these movements, but
crude totals tended to obscure a significant out-movement,
especially of better-educated school-leavers (Kayser, 1973).

Lévy's (1973) study of Auch, chef-lieu of the département
of Gers, gives a useful insight into the nature of population
and employment change in a rather isolated ville moyenne.
Half the active population of Gers was still engaged in agriculture in 1968 and the département has been losing population
for more than a century. In Auch itself, however, this
pattern of population loss was reversed after the First World

War and the town experienced an annual growth rate of 2.2%
between 1962 and 1968 (Figure 9).   The fact that 84% of this
growth was attributable to migration points to the importance
of rural to urban movements, but it is noteworthy that the 1968
population total of 21,462 also includes 1,058 repatriates
from North Africa and 1,286 foreigners, mainly Algerians and
Portuguese, the latter engaged principally in construction and
labouring occupations.

   The main components of migratory movements in the 1960s
are further illustrated in work on Montauban (Idrac, 1973) and
Albi (Bienfait, 1973).   Montauban (1968 total: 45,895)
experienced a net gain from migration of over 3,000 between
1962 and 1968, but the migratory 'flux' was very much greater
with no fewer than 11,544 migrants entering the town over this
period and 8,358 leaving.   The arrivals were made up as
follows:

Tarn-et-Garonne	3,032
Remainder of Midi-Pyrénées	1,948
Remainder of France	3,272
Repatriates from Algeria	2,152
Foreigners	1,140

The migrants from within the département, and probably many of
those from the region, were mainly young, unmarried and of
rural origin, different in most respects from the repatriates
and those from other parts of France, including Paris.   The
latter were older, typically civil servants with their families,
coming to take up positions in the public services.   Some were
near the end of their career, returning to their region of
origin.   The outward migrant tended to be young, educated in
Montauban but compelled to seek work elsewhere and included
more males than females.   The scale of this outward movement
led Idrac to comment on the lethargy of the local economy.

   The picture was similar in Albi (1968 total in the agglomération: 53,365) where 11,876 migrants entered the town between
1962 and 1968, but over 7,000 persons left it.   A large proportion of the latter were aged 15-34 and were moving to the
cities (including 1,176 to Toulouse).   Some were students
seeking higher education but others were moving in search of
employment and this, too, is indicative of the town's failure
to provide a sufficiently wide range of work to retain its
school-leavers.

   Heavy dependence on a relatively narrow range of jobs in
the tertiary sector is brought out in all the studies.   No
less than 67.6% of the working population of Auch was employed
in this sector in 1968, a large proportion in administration,
education, health and other public services.   Lévy refers to
the 'tertiarisation de l'économie' and describes Auch as
increasingly 'une ville moyenne pour classes moyennes' on
account of its attraction for white-collar public servants.

   Lévy and Poinard (1973) emphasize the importance of

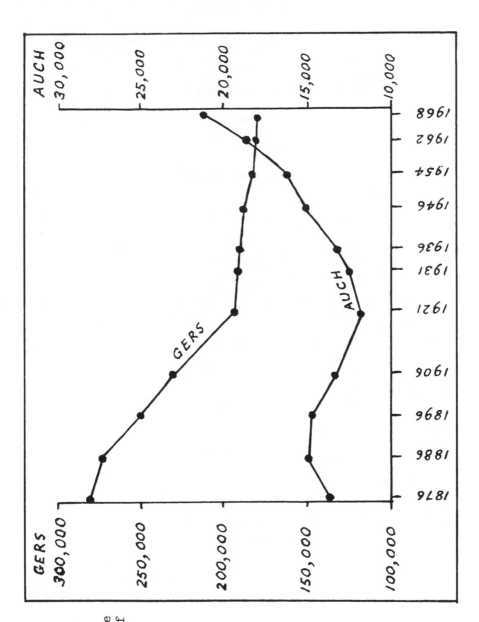

Figure 9.
Population change
in département of
Gers and town of
Auch

services in the population growth of all <u>villes moyennes</u> in
the south-west, none having less than 50% of their employed
population in service occupations and the proportion reaching
70% in a few cases.   Expansion of the public sector accounts
for the largest single share, but this in turn has acted as a
stimulus to organizations engaged in finance and commerce
(superstores, car sales and other forms of distribution) and
in construction.   The authors observe that growth of white
collar employment has contributed, however, to a rise in the
cost of living which in turn has encouraged industry, where
it is present, to rely to an increasing extent on cheaper
foreign labour.   They conclude that the growth of the tertiary
sector has brought a certain prosperity to the towns but has
made them more vulnerable, for example, to the effects of cuts
in public services.   This conclusion is echoed in Idrac's
description of Montauban as 'une ville à l'économie fragile'
(Figure 10) and Tulet's (1973) comment that the growth of
Cahors was 'croissance sans développement'.

It was believed by some (e.g. Prager, 1973) that an
improved level of service provision would, itself, act as an
inducement to manufacturing industry which would be drawn to
the medium-sized towns in order to take advantage of the
better financial, educational and other facilities ('l'induction
du tertiaire').   Many of the towns had, in fact, laid out
industrial estates during the 1960s in the hope of attracting
firms to them and some had been successful.   But the pattern
of success had been a very uneven one, the spread of decentralization having benefitted principally the medium-sized
towns of the Paris Basin and those close to Lyon and Marseille.
The more distant places in, for example, the south-west had
gained little from government policies except in a few cases
where local circumstances, political or personal, appear to
have acted as a special inducement.

According to Michel (1977) the attraction of a medium-
sized town to an employer in the 1960s or early 1970s lay in
its supply of relatively cheap labour, but this pull was
exerted only within a limited radius of the major cities:
' ... l'existence d'une importante force de travail sous-
employée, dans une aire suffisamment proche de la ville, crée
un atout potentiel que l'économie industrielle ne peut manquer
de mettre à profit'.   Populations swelled by migration from
the countryside and abroad provided a useful workforce at a
time of rapid industrial expansion and the incoming firms were
generally welcomed by the local authorities concerned.   But
their arrival was not an unmixed blessing and Michel draws
attention to two resultant sources of weakness.

The first was a general lowering in the overall standard
and level of qualification of the industrial workforce which
followed from the fact that most of the migrant firms were
concerned more with numbers of employees than with skills and

70

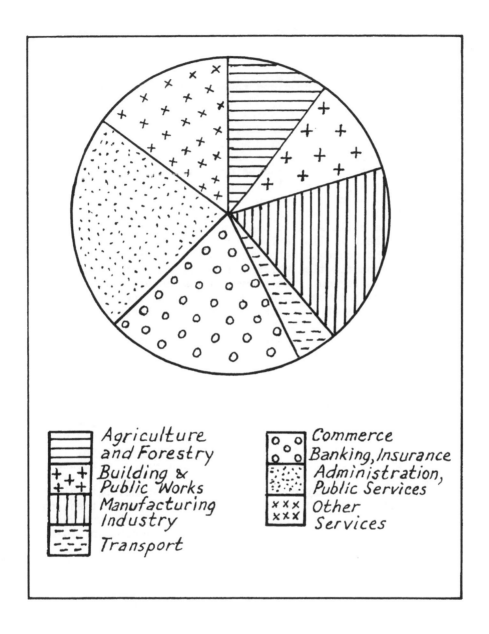

Figure 10. Employment structure of Montauban, 1968

that the kind of jobs being created were of a poorly-paid, repetitive and undemanding nature. 'Pour l'instant, la croissance des "villes moyennes" a conduit à leur prolétarisation' (Michel, p.679). Lajugie (1974) refers to the danger of the medium-sized towns developing a 'sous-qualification économique' (p.63) in spite of their apparent demographic buoyancy. He notes that in 1968, 43% of the active population in towns of under 200,000 was blue collar (personnel ouvrier) compared with under 38% in those exceeding 200,000. The weakness of trade union organization in the smaller towns usually meant that little was being done to improve the conditions of this poorly-skilled workforce.

A second weakness lies in the continued concentration of decision-making in Paris or the other large cities. Decentralized firms still look to the capital for direction and advice and there is little local responsibility for their affairs. The consequence of this external orientation is to loosen the economic ties which the medium-sized towns have with their own region. In these circumstances, far from creating a new balance in the urban system at the regional scale, the effect of decentralization is to reinforce, even disrupt, existing structures.

Awareness of these economic weaknesses, as well as of the social problems resulting from rapid population growth, was the context in which planners and politicians sought to formulate a politique des villes moyennes.

A Policy for the Medium-sized Towns

Government support for the medium-sized towns was expressed in March, 1972 when it was announced that pilot projects were to be carried out in up to half-a-dozen villes moyennes. Experience in these would determine the pattern of a future policy favouring such towns. In fact only one town, Angoulême, was actually chosen for a pilot scheme, but this does not seem to have affected the government's commitment to the policy, details of which were set out by M.Guichard when he addressed the CNERP conference at Nice in October of 1972. The following February, M.Guichard directed a letter to all the departmental préfets in which he stated that it was now possible for individual town councils to enter into a contrat d'aménagement with central government. At the same time he announced the formation of a Groupe Opérationnel des Villes Moyennes made up of representatives from DATAR and DAFU (Direction de l'Aménagement Foncier et de l'Urbanisme), the object of which was to offer advice and technical assistance to those local authorities seeking to enter into a contract. Money was also set aside for financing the work. The politique des villes moyennes had been launched.

Before looking in detail at the nature of the contracts entered into, it is necessary to establish what is meant by a medium-sized town.   For official purposes this term embraces all agglomérations with a population of between 20,000 and 200,000 and, defined in this way, France had 193 villes moyennes at the date of the 1968 census.   Between them they housed just over 11 million people, nearly a third of the country's urban population, and their size-distribution was as follows:

Population	20/30,000	30/40,000	40/50,000	50/100,000	100/200,000
No. of towns	65	28	26	46	28

Approximately 3,750,000 people lived in towns of between 20,000 and 50,000 (10.5% of the total urban population); 3,250,000 in those of 50,000 to 100,000 (9.5%); and some 4 million in towns of between 100,000 and 200,000 (11.5%).

The 193 villes moyennes exhibited considerable diversity, not only of size, but also of geographical location.   DAFU recognized five types:

1.   Towns forming part of a larger urban region.   Examples quoted include Versailles and Saint-Germain-en-Laye, essentially suburbs of Paris, Longwy in Lorraine, and Armentières in the northern industrial conurbation.

2.   Towns that are satellite to a larger city, being influenced strongly by the latter.   Examples are Mantes and Melun in the case of Paris, Vienne and Givors close to Lyon, and Elbeuf near Rouen.

3.   Free-standing towns such as Rodez, Auch, Tulle and Draguignan.   Fulfilling the role of local service centre and hopefully of centre relais, these are described as the 'archetypal' ville moyenne.

4.   Towns that fall between categories 2 and 3 in the extent to which they are influenced by large centres of population.   Examples are Roanne (Lyon), Arras (Lille), Chartres (Paris) and Haguenau (Strasbourg).

5.   Towns that owe their importance to local resources or the presence of some related advantage.   This category includes coastal resorts such as Royan and Les Sables-d'Olonne, spas (Vichy, Thonon), towns serving the military (Rochefort), ports (Sète) and, most commonly, centres of mining or manufacturing (Montceau-les-Mines, Forbach, Le Creusot, Carmaux, Cluses).

Commenting on this list, Lajugie (1974) felt that there was some justification for including a sixth category of

ville moyenne, the closely associated groups of small towns
found in areas such as Brittany.   He described these as 'une
grappe de villes' on account of their functional interdepen-
dence.

Even without this addition, it is clear that the term
ville moyenne covers a wide range of urban types and it would
be an over-simplification to suggest that there was a single
model or prototype of the medium-sized town.   This fact was
recognized by the authorities who were at pains to point out
that individual contrats d'aménagement should be tailored to
the needs of the particular town concerned.   Government state-
ments, together with Lajugie's report of 1973, suggest that
the scope of the politique des villes moyennes was seen as
wide-ranging and many different kinds of action envisaged.
These might include the search for new forms of employment,
provision of housing and related services, and the improvement of transport and communication links with both the
regional metropolis and with the town's own rural hinterland.
Attention should also be given to the quality of educational
and cultural facilities and to the role which the town could
play in the revival of the surrounding countryside.

In drawing up their list of projects to be included in a
contract, local councils were urged to adopt a comprehensive
approach to planning and to avoid so far as possible the frag-
mentation of decision-making that had resulted in the slow
adoption and subsequent abandonment of many earlier plans.
The problem of plan-making was highlighted in a study of five
medium-sized towns (Aix, Annecy, La Rochelle, Poitiers,
Valence) that had been carried out by the Centre de Recherche
d'Urbanisme and was made the subject of a conference organized
by the CRU in 1974 under the title of 'Trente ans d'expérience
de la planification urbaine dans cinq villes'.   The investi-
gation revealed that of a total of 48 projets d'urbanisme, no
fewer that 31 had been abandoned, and that the five towns had
had the benefit of an approved plan to direct their urban
development for an average of only eight out of the preceding
30 years.

Further clues to what government officials had in mind
may be obtained from an examination of the first three contracts to be drawn up in 1972-73, with the municipalities of
Rodez, Angoulême and Saint-Omer.

Rodez, though not formally designated ville moyenne pilote,
did in fact serve as a pilot scheme and when M.Guichard sent
his letter to the préfets in 1973 he included with it a copy
of the agreement that had already been drawn up with the town
council of Rodez.   This was clearly intended to serve as an
example, even a blueprint, of what could be done.

Administrative and service centre of the département of
Aveyron, Rodez is an undisputed local capital, comparable in
many ways with other towns of the south-west described above.

The Aveyron's population had fallen from a total of 415,000 in
1886 to only 281,000 in 1968, but that of Rodez itself had
grown to reach 29,952 at the latter date.   A high proportion
of the town's working population were engaged in the service
sector (68%) and the growth of employment in public services
had been important here as elsewhere.   There are several
lycées and colleges and, in addition, Rodez has the offices
of several prominent organizations that serve the agricultural
needs of the surrounding area.   These organizations have
played a major part in the transformation of agriculture that
has taken place in Aveyron since the Second World War.   It
has been established on a much more commercial basis and the
town now houses the headquarters of various unions and
cooperatives as well as factories which supply feedstuffs and
machinery or manufacture food products.   There was benefit
to Rodez in having as mayor a former Minister of Agriculture
who also served as Vice-President of the National Assembly
between 1968 and 1972.

     Employment in manufacturing was limited, except for that
connected with agriculture.   An exception was the firm of
Bosch, employing around 1,000 workers in the manufacture of
injection pumps.   This enterprise was established in Rodez in
1939 when M.Bessières, an Aveyronnais, moved his precision
engineering company from Paris, and although taken over by the
German company in 1960, it has continued to expand.   In the
early 1970s the example of this firm was quoted as proof both
of the fact that manufacturers with regional connections might,
given sufficient inducement, be tempted back to the area, and
that a firm which was engaged in the production of specialized
and high-value products could prosper in spite of the distance
from major cities (Rodez is 610 km from Paris and 160 km from
Toulouse).   Two industrial estates have been laid out in the
effort to make Rodez more attractive to manufacturers.   A
Class 'C' airport has been built, and pressure brought to bear
on the authorities to improve the town's external road connections (Lugan and Poinard, 1973).   Several small firms have
been drawn to the town, but it is probably an exaggeration to
claim that, 'A generation of urban ex-peasants crave a return
to their local roots, not of course to the farms, but at least
to smaller factories or offices near the home patch'
(John Ardagh in The Sunday Times,12 September, 1976).

     'Dix actions sur le centre' was the title of the contrat
d'aménagement drawn up for Rodez (nine operations were, in
fact, agreed).   As that title suggests, the emphasis was on
the town centre, enabling the older parts of the town to
adjust to the changed circumstances of the 1970s.   Amongst
the proposals accepted were plans to close selected streets
to traffic, to widen pavements and plant trees, to restore
historic buildings and make provision for galleries, theatre
clubs and street cafes, to encourage arts and crafts, and to

install a new system of street lighting that was in keeping
with the character of the old town. The whole programme was
expected to take five years to complete (it required seven)
and to cost 14.2 million francs, 45% of which would be borne
by the state. The final cost was, in fact, slightly in
excess of 20 millions but the following observation by the
Mayor of Rodez (in Le Moniteur, 1981) suggests that the money
had been spent profitably: 'La vieille ville se réveille,
nous sentons son coeur battre et cela est fort agréable.
Rodez ... a su conserver une âme'.

The plan to carry out pilot studies seems to have foundered on the political problems arising from the selection of
up to half-a-dozen towns to benefit from the scheme. The
retention of Angoulême as ville moyenne pilote was seen by
some as a compensatory gesture to make up for the disappoint-
ment felt in the town when it was decided that the A.10 motorway from Poitiers to Bordeaux should pass through Saintes and
not through Angoulême (Comby, 1973). Even here the formulation of a strategy was slow. The decision to go ahead with
a pilot scheme in Angoulême was taken at the end of March
1972 but it was October 1973 before the details of the contract were finally agreed and signed.

Fourteen operations were included in the dossier
opérationnel (Comby, 1974). A schéma directeur d'aménagement
et d'urbanisme (SDAU) for Angoulême and its surrounding
communes had, in fact, been adopted during the course of 1972
and the government, for its part, made it clear that the
contrat ville moyenne should complement this development plan,
adding a gloss as it were to projects that were already in
hand. Thus half the operations proposed were concerned with
roads or the provision of car parking. They included a new
bridge to be built across the Charente in connection with a
projected inner ringroad, and an improved road link between
a major new housing development and the town centre.
Prominent amongst the rest of the proposals were plans for
improving the appearance and the quality of services available
on suburban housing estates. A government minister
(M.Chalandon) had been particularly critical of one of these
large ZUPs (zone à urbaniser en priorité) which he had described as 'concue de manière trop traditionnelle' (Le Monde,
14 April, 1972), and the contract referred to a need (to)
'humaniser' the area. This was to be done through the provision of sports facilities, nursery and medical centres,
banks and post offices, and by works of an environmental
nature that would include tree-planting and the protection of
open spaces.

It is evident from the above that the emphasis in
Angoulême's contrat d'aménagement was different from that of
Rodez. It reflects the problem of a larger town which was
under some pressure to house and service its growing popu-

76

lation and which was experiencing problems of traffic circulation, due partly to this growth and partly to the local circumstances of an accidented site. It reflects, also, the local political situation. The population living within the municipal boundary of Angoulême was slightly less than 50,000 in the early 1970s but the town council had entered into negotiations with the surrounding suburban communes to set up a syndicat intercommunal à vocation multiple in order to coordinate planning measures. The population of the whole area was some 100,000 and without this cooperation it is unlikely that Angoulême's contract would have placed so much emphasis on the outer parts of the town. There was a lesson here for other municipalities, particularly since the government had expressed its strong wish that contracts should be tailored to meet local needs. But, as we shall see, the failure to achieve intercommunal cooperation resulted in many of the contractual agreements concerning themselves largely, even exclusively, with the central commune and its historic core. An exception to this generalization was the contract agreed between Saint-Omer and the government in December, 1973.

Lying almost midway between the coast and the towns of the northern coalfield, Saint-Omer was little touched by the industrial revolution, and preserved a genteel, bourgeois atmosphere. But change has accompanied the development in recent years of industries, notably glass-working and the manufacture of telephone equipment, and the famous hortillonages - areas of marshland cultivation of the Aa valley - have been threatened by the further pressures that have followed the completion of the deep water canal from Dunkerque. The need to accommodate this growth was recognized as early as 1962 when the first steps were taken to create a district urbain. The latter now takes in 18 communes with a total population expected to reach 80,000 by 1985 and possibly 100,000 by the end of the century. Saint-Omer's contrat d'aménagement acknowledges this involvement of the town with its surrounding area and the thirteen operations that were agreed include measures to preserve 7,000 hectares of marshland. Others of a more conventional nature involve the creation of a park on the slopes of Vauban's ramparts, restoration of a seventeenth-century barracks, creation of a new library in a disused chapel and various traffic measures.

A Popular Strategy

Town councils were swift to recognize the possibilities afforded by the government's promotion of the villes moyennes. Although the strategy was intended to assist provincial towns, the suggestion was made by the préfet of the Paris Region that it might even be extended to the planning of the Ile-de-France. Parallels were drawn between the role of the ville moyenne in complementing the métropole d'équilibre and the contribution which places such as Nemours, Fontainebleau and Coulommiers might make to balance the five new towns designated under the Schéma Directeur for Greater Paris. Their populations, it was suggested, could be doubled by 1990 without loss to their essential character. The idea was not well received in DATAR, however, where attempts to promote the interests of Paris at the expense of the provinces were always viewed with suspicion.

Elsewhere, enquiries were received more sympathetically, and by March, 1976 the number of towns seeking a contrat d'aménagement had grown to 82. Of these, 28 had already obtained a contract, 23 were awaiting final confirmation of contracts that had been drawn up for agreement, and 31 others were still negotiating with the authorities over the details of a possible contract. In a study published in Le Monde (17 March, 1976) Michèle Champenois looked at the nature of these 82 contrats, agreed or proposed, to see whether there had been any shift in emphasis in the strategy since 1972 when the first suggestions had been put forward. Since the schemes were intended to meet local needs, any such shift was not expected to be great. Nevertheless, Champenois observed some slight change over the four years. There had been a reduction of emphasis over this time on environmental improvement projects, including the creation of open spaces and pedestrianized streets - what critics described as the 'pots de fleurs' approach - the expenditure set aside for which had fallen from 46% of the total in the earlier schemes to 27% in the later ones. There had been a corresponding growth in spending on house improvement and on the provision of social and cultural amenities. Expenditure on road building and car parking had similarly fallen from 19 to 13% of the total, greater interest by 1976 being shown in traffic management measures.

Outside the Paris Basin the distribution of the 82 towns was fairly even. It is likely that politics played a part in the determination of those places chosen to benefit from a contract, but this is difficult to assess. Of the 82 municipalities involved in negotiations, 58 had a town council of a similar centre-right persuasion to that of central government, but this apparent bias can be explained in part by the disinclination of some left wing councils to appear to be seeking favours of the authorities in Paris.

March, 1976 marks a turning point in the politique des villes moyennes because the government now made it clear that it would not welcome any additions to the list of towns seeking contracts. Instead, town councils were urged to take advantage of the newly established Fonds d'Aménagement Urbain in order to finance their urban improvement schemes. Negotiations with the 82 towns already on the list were not broken off, however, and contracts continued to be signed until 1979. In a few cases the negotiations proved fruitless and the final total of towns to benefit was 73 (Figure 8).

Most of the towns awarded contracts had a population of under 100,000, and in some cases little more than the 20,000 regarded as the minimum needed to qualify as ville moyenne. Many were chefs-lieux of their respective départements, or sous-préfectures, and, like the three described above, all were seeking to adjust to the pressures brought about by population growth.   Vesoul (population 21,000), préfecture of Haute-Saône, provides a good example, although the scope of its contract ('Faire de Vesoul une ville', 1976) was more ambitious than most.   It was unusual in listing an economic objective first - 'la politique d'économie qualitative' - a reflection perhaps of the town's dependence in manufacturing on a single firm (Peugeot).   It also sought to create a major outdoor sports complex based on an artificial 70 ha. lake.   This was to account for 65% of total expenditure.   Charles (1979) compares Vesoul's contract with that drawn up for Dole (30,500), historic capital of Franche-Comté, where the emphasis was firmly on restoration and renewal in the old town.   The aim of establishing walks and viewpoints and of finding new cultural uses for historic buildings, was intended to complement the programme of repairs and conservation drawn up when a secteur sauvegardé had been designated in 1967.

It would be difficult to find a better example of the problems of reconciling the old France and the new than that provided by the town of Manosque in the southern Alps.   The old town is perched well above the Durance, its tightly-packed houses built above the stables which gave shelter to the flocks of sheep moving seasonally between the plains and the mountains.   This is 'Manosque des Moutons'.   But in the valley below there are hydro-electric plants and the atomic research station of Cadarache, and these have attracted a new population of power workers, scientists and their families. Repatriates and immigrants from North Africa have swelled the total which has grown from around 5,000 to 20,000 within a generation.   A third of the population is of school age, and the need to provide housing and services for the newcomers has led to the creation of a new settlement extra-muros.   This is 'Manosque de l'Atome' (Rambaud, Le Monde, 17 March, 1976).

One of the effects of this new development has been to draw the better-off and more mobile sections of the established population away from the old town towards the facilities of the newer part, abandoning the crumbling quarters of the former to the elderly, the poor and the immigrants.   Old houses become what Rambaud describes as taudis historiques - historic slums. The challenge posed by Manosque is similar to that encountered in many French towns, though experienced here in a rather extreme form, and the aim of the contrat ville moyenne has been to save the old town from further decay with a programme of works that combines physical repairs with an attempt at reanimation and improvement of the 'quality of life'.   To this end _foyers de vie_ have been installed in some of the historic buildings following their repair. One of them now houses the municipal library, another a school of music, whilst others have been used as council offices or altered to accommodate the elderly. Overall the aim has been to revive the old town and make it an attractive environment in which to live and work.

Much the same could be said of the projects carried out in the larger town of Périgueux, _chef-lieu_ of the _département_ of Dordogne, where all twelve proposals agreed in the contract of 1977 refer to the central parts of the agglomeration. Modern Périgueux has grown around two historic nodes, a Roman town, the site of which is still known as La Cité, and a monastic suburb where fine houses of the fifteenth, sixteenth and seventeenth centuries line the narrow streets converging on the domed cathedral of Saint-Front (Scargill, 1974). A route centre on the river Isle, the town has always acted as administrative and commercial focus of its region, first of the _comté_ of Périgord, later of the _département_. Services employ two-thirds of the working population and expansion of these accounts for most of the growth in employment that has taken place, manufacturing being largely confined to non-growth sectors, notably railway engineering, textiles and food trades. An industrial estate opened in the 1960s has been slow to attract new firms, although the printing of French postage stamps was transferred to Périgueux in 1970 when the town's mayor was Minister of Posts and Telecommunications.

The Périgueux agglomeration recorded a population of some 60,000 in 1975, but the central commune of Périgueux itself accounted for only 37,000 of that total, having lost population to the suburban communes at an annual rate of 0.5% between 1962 and 1968 and by 0.8% a year 1968-75. It was to the related problems of decay at the centre and the accommodation of growth on the periphery that the local authorities addressed themselves when preparing submissions for a _contrat ville moyenne_ in the early 1970s. A _dossier d'intention_ was sent to the Groupe Interministériel des Villes Moyennes early in 1975. It was more than two years later before details of the contract had been agreed, however, and during that time some significant changes had been made to the original set of proposals. In particular all plans for the outer part of the agglomeration had been dropped, investment now being concentrated on the central commune. The contract that was eventually signed in 1977 referred to 'la consistance nouvelle du programme', but its critics were more inclined to attribute the changes to the longstanding rivalry between _la ville_ and its peripheral communes. The mayor of Périgueux was a gaullist and strong supporter of the government whilst the suburbs were mainly represented by left-wingers, in some cases members of the _parti communiste_.

The agreed contract contained 12 sets of proposals, all

aimed at sustaining the role of the centre and at improving living conditions there (Périgueux: Contrat Ville Moyenne, 1977). Details are set out below and the location of the projects can be seen in Figure 11.

	Proposal	Cost (francs)
1.	A new bus station in the Place Francheville	5,116,000
2.	Environmental works in the vicinity of the cathedral	2,237,000
3.	Extension of works already in progress in the 'old town' (of Saint-Front) involving paving, removal of overhead wires, etc.	9,608,000
4.	Creation of an open space by removal of the old fire station	428,000
5.	Restoration of historic buildings in one îlot of the 'old town'	547,000
6.	Road-widening at one of the approaches to the 'old town'	555,000
7.	Extension of the Hôtel de Ville	250,000
8.	Restoration of a seventeenth-century convent for use in connection with cultural activities	590,000
9.	Preliminary work on the site of a former old people's home	180,000
10.	Conversion of a former municipal depot	40,000
11.	Establishment of a Gallo-Roman museum on the site of an excavated villa	200,000
12.	Improved access to the restored convent	50,000
		19,801,000

This programme of works was largely completed during the three years, 1978-80. It may be regarded as typical of many of the schemes carried out in villes moyennes elsewhere in France. The largest single item of cost was for the restoration of buildings and environmental improvements in the historic core of Saint-Front and this may be seen as an extension of the work begun there after the whole of the 'old town' had been designated a secteur sauvegardé in 1970. Two secteurs opérationnels had subsequently been defined. In one of these the emphasis was on public works: modernizing the water supply and drainage system, putting electricity and telephone wires underground, and repairing the narrow streets in a traditional manner using pebbles from the river and stone slabs for paths and gutters. In the other one, old property had been demolished to make way for a hotel, nursery school and new housing built above an underground car park. The latter had proved controversial on account of the demolition involved, and the stress in the contrat ville moyenne was on renewal rather than redevelopment. It can be seen from the

82

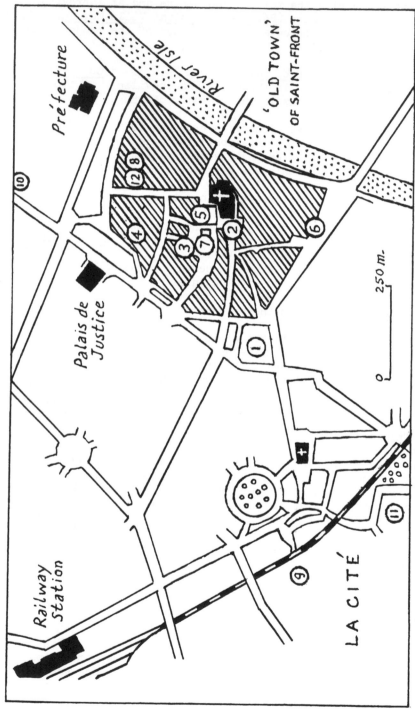

Figure 11. Périgueux: location of the 12 projects carried out under the 'contrat ville moyenne'

list of proposals above that very nearly three-quarters of the total expenditure in Périgueux's contrat ville moyenne involved additions to this work of restoration in the 'old town'.

Preoccupation with the core has been regarded as one of the failings of the politique des villes moyennes. It tends to divorce the problems of the centre from those of the modern suburbs where an increasing proportion of the population now lives, paying little, if any, attention to the social needs of the ZUPs, the conflicts of land use resulting from commercial pressures on the fringe, and the difficulties of traffic circulation throughout the urban area as a whole. Thus, in Périgueux, for example, there is no attempt in the contrat to resolve the longstanding problem over the route of an inner relief road. Supporters of the politique argue, however, that these wider planning issues are more properly the concern of the town plan (SDAU) and that the object of the contrat is to complement existing projects, not to provide an alternative to them. One writer sees the contract as contributing 'soul' to the town plan: 'Le contrat de ville moyenne ajoute en quelque sorte un "supplément d'âme" a l'aménagement prévu antérieurement' (Charles, 1979). Interpreted in this way, the principal achievement of the short-lived politique des villes moyennes probably lies in the public response it has evoked to the needs of the historic provincial town. The advice and financial help available from Paris has, in many cases, acted as a catalyst to change, encouraging rather conservative local authorities to see their old buildings in a new light and to respond to their needs.

The Small Towns

The 1970s were marked in France, as in other western industrial countries, by a growing appreciation of the difficulties faced by rural areas. Loss of population, the reorganization of agriculture, contraction of services, and conflicts of interest resulting from the growth of tourism and second home ownership, were amongst the main causes of these problems. There were references to the dévitalisation of the countryside and serious concern was expressed for the future of la France fragile.

Amongst the strategies evolved for assisting the country areas was that of the contrat de pays, an arrangement by which a number of rural communes could draw on government, later regional, funds in order to carry through a programme of works aimed at creating new forms of employment and at improving the level of local service provision. A condition of the contract, however, was that the communes concerned should be ones that looked to a small town as centre, the idea being that co-ordination and co-operation could best be achieved if it were directed from a recognized central place, usually the local market town. 'Il s'agit d'une "formule" de coopération

communale, en zone rural, qui permet en général aux villes
chef-lieux de cantons ou d'arrondissements de nouer des liens
de solidarité avec les petites communes voisines' (Le Monde,
3 March, 1980).   In the emphasis which it placed on the role
of the small town, the politique des contrats de pays (referred
to also as 'la politique des petites villes et de leur pays')
can be regarded as an extension of the politique des villes
moyennes, extending to the lower levels of the urban hierarchy
the kind of assistance that had earlier been reserved for the
medium-sized towns.   Some went so far as to see it as the
third and final stage in the restructuring of the urban system
which had begun with the métropoles d'équilibre and been continued with the villes moyennes.

Details of the new strategy were announced by the government in 1975 and twelve schemes were launched on an experi-
mental basis in that year.   One of these was at Loudun in
Poitou where 52 communes were involved, and five years later
the mayor of Loudun was able to claim that the contract had
reconciled townsmen and country-dwellers, 'Ce contrat a permis
aux communes rivales de retrouver l'espoir.   Il a permis
aussi d'affirmer la solidarité entre les Loudunais et les
ruraux' (from a speech to the first Journée Nationale des
Contrats de Pays held at Poitiers, 29 February, 1980).

To help promote the strategy, DATAR published in 1976 a
study of the 533 towns in France with a population of between
five and twenty thousand (La Documentation Française, 1976).
Whilst noting that their population had risen at the annual
rate of 1.2% between 1968 and 1975, the report also drew attention to a number of structural weaknesses.   A slightly higher
proportion of their working population was employed in manufacturing industry than was the case in other towns (36% compared with 33%) but for many of them this meant heavy dependence on a single category of manufacture, often on one firm.
There had been growth in service employment in the small towns
as in the larger ones, but this tended to be limited to the
less well paid sectors and those carrying least responsibility,
and in the case of commerce the benefits of growth risked
being lost to competition from the out-of-town hypermarkets
and superstores.   In his introduction to the report,
M.François Essig, who had succeeded M.Monod as head of DATAR,
emphasized the contribution which the contrat de pays could
make to the future of the small town as well as to its surrounding communes.   'La situation des petites villes est
encourageante, mais reste fragile.   D'où l'intéret de la
politique contrats de pays.'

Fifty-one contrats de pays were signed during the course
of 1976 and by 1980 the number had risen to more than 280,
affecting 7,500 communes and a total population of some 5
millions.   During this time there was a gradual shift in the
responsibility for financing the agreed programmes from the

state to the regions.   The 'reforms' of 1973 which created
the établissement public régional had given to the regions
rather greater authority in matters of planning and also the
resources to pay for modest developments.   The first of the
regions to take advantage of these new powers was that of
Centre in 1975.   Three others followed in 1976 and by 1978 no
fewer than fifteen of the EPRs were entering into contracts,
in this case known as contrats régionaux d'aménagement rural,
although a part of the overall cost continued to be borne by
central government.

In addition, a number of the regions also decided to
copy the politique des villes moyennes and extend to selected
small towns some of the benefits that had earlier been
reserved for those of medium size under the national strategy.
Agreements known as contrats des villes moyennes régionales
were signed, the intention being that such a contract might
benefit the small town alone or might be combined with a
contrat de pays.   It was justified on the grounds that a
population total of 20,000 was an arbitrary distinction between
what constituted a small or a medium-sized town and that population alone was in any case an inadequate measure of the
status and needs of a particular urban centre.   The termi-
nology attracted some criticism, however, some arguing that
the title ville moyenne régionale was no more than a euphemism
for petite ville, used as an alternative to the latter in
order to avoid offending the sensibilities of local politicians
(Lajugie et al., 1979).   The term, ville d'appui, employed by
the EPR of Bourgogne may be thought more appropriate given the
small town's role as rural service centre.   But it was not
always clear how this supporting role was to be exercised and
Charrier (1977) quotes the example of Decize (1975 population:
9,700) in Burgundy where the contrat de ville d'appui signed
in 1976 was similar to a contrat ville moyenne in its stress
on central area restoration and improvement.

The small town of Ribérac in Dordogne may be regarded as
typical of the many that have profited from this kind of
strategy.   Although its population in 1975 was no more than
3,984, Ribérac is a lively market town serving a well-defined
pays (the Ribéracois) in the basin of the river Dronne some
35 km west of Périgueux.   Its contract, with the Région of
Aquitaine, was signed in 1978 and involved two projects.   The
first was an extension to the market hall so that it might
serve as a place for public meetings, concerts and entertain-
ments as well as satisfying the needs of the local traders.
The second involved the complete remodelling of the central
town square and adjacent public garden in order to improve
their appearance and to exercise greater control over car
parking.   The anticipated cost was 2.1 million francs to
which the Région agreed to contribute 800,000 francs, 38% of
the total.   Work on both projects was carried out in 1979 and

1980. At the same time the town was also engaged with neighbouring communes on a number of small schemes agreed in connection with an earlier contrat de pays.

In Retrospect
It was one of the intentions of the politique des villes
moyennes that the medium-sized towns should be helped to perform what Lajugie (1974) describes as 'leur rôle de relais du
développement' (p.149). In retrospect one can see little
connection between the choice of towns for contracts and the
objectives of regional planning. The quantitative term,
ville moyenne, itself conveys no idea of relationship to territory, and there is little reason to think that policies for
the medium-sized and small towns have themselves led to any
significant hierarchical diffusion or spread of growth
throughout the country or to the establishment of a new set of
vertical and horizontal linkages within the national urban
system. Change within the system has undoubtedly taken place,
but growth has been uneven in space, and appears to owe most
to proximity to the major cities. With a few exceptions, the
most buoyant of the villes moyennes have been those in the
Paris Basin, the Rhône-Saône valleys and the middle and lower
basin of the Loire. The introduction of the contrat de pays,
leading to the involvement of the regions in the planning of
smaller towns, has shown a greater awareness of the links
between town and country, but at an essentially local scale.
Arguably what is required in the 1980s is a strategy for the
regions that recognizes the complexity of association existing
between towns at the regional scale. Representatives of DATAR
have referred to this as a politique des aires urbaines
(Bouchet and Muron, 1978). Noin drew attention to the regional function of the major urban centres in L'Espace Français
and to the potential of these regions for territorial planning. The implementation of decentralization plans, begun
in 1982, may result in an awareness of the links between
towns that, for the most part, has been lacking from urban
strategies over the last twenty years.

Chapter 5.

HOUSING AND LAND USE PLANNING

A survey of French housing carried out for INSEE in the summer of 1978 revealed a total of 22,235,700 dwellings of which 18,641,000 (83.8%) were recorded as occupied and the principal place of residence of a household. Second homes accounted for a further 1,843,500 (8.3%), whilst 1,751,100 (7.9%) were vacant.

Of the 18,641,000 primary residences, almost half (9,083,000 = 48.7%) had been built since 1949. Not unexpectedly, this proportion is higher in urban areas than in the countryside. It was 51.3% in the Paris Region, rising to almost 60% (59.9%) in the other large agglomerations (all those, except Paris, having a population in excess of 100,000).

Flats or apartments accounted for just over 8 million (43.2%) of the primary residences, the remainder being classed as individual dwellings. This latter category, however, embraced a wide diversity of house types ranging from farmsteads and manors to suburban pavillons and terraced houses of the nineteenth century like the corons of the industrial Nord. Collectifs, i.e. flats and apartments, predominate in the larger urban centres, accounting for no less than three-quarters (75.5%) of the total housing stock in the Paris Region. The proportion was almost 60% (59.5%) in the other large agglomerations.

The amount of living space declines with the emphasis on collectifs as a form of housing. The 1978 survey found the average number of rooms per dwelling in France to be 3.7, but there was a wide difference between the average size of flats and apartments (3.0 rooms) and that of individual dwellings (4.2 rooms). Living space likewise ranged from 63 sq.m in collectifs to 88 sq.m in individual dwellings, the national average being 77 sq.m. It follows from the emphasis on collectifs in the cities that the amount of living space per dwelling is lowest there. The average for all dwellings in the Paris Region was only 3.1 rooms and 65 sq.m of living space; in the other large agglomerations the figures were

closer to the national average (3.6 rooms and 75 sq.m).
    The figures below show the proportions of the primary
residences that were owned or rented in 1978.

Owner-occupied	8,695,033	= 46.7%
Rented unfurnished	7,652,416	= 41.0%
Sub-let or rented furnished	372,051	= 2.0%
Let free	1,921,560	= 10.3%

Renting increases in importance in the cities. Little more
than a third (35.8%) of the dwellings in the Paris Region were
owner-occupied in 1978, and the proportion was not very much
higher (38.3%) in the other large agglomerations. The total
of properties rented unfurnished included 2,481,000 dwellings
(13.3% of all housing stock) rented from the agencies responsible for HLM housing (habitations à loyer modéré).

Legislation for Housing
    The housing construction industry was one of the most
depressed sectors of the French economy between the wars.
The slow rate of growth, both of population and of production,
did not encourage new building, but the principal explanation
lay in legislation introduced in 1914 in order to protect
tenants. Rents were frozen in August of that year as a wartime measure and although a number of controlled rises were
permitted during the interwar years, no effective attempt was
made to relax the rigid limitations that had been imposed
until after the Second World War. Tenants enjoyed not only
security of tenure but also a hedge against inflation and it
has been estimated that by 1948 the average Parisian family
renting a property was devoting only 1.5% of its family budget
to rent (INSEE). In such a situation there was little incen-
tive to invest in new housing, whilst existing property was
neglected by landlords unable or unwilling to carry out repairs
and improvements.
    Equally strong deterrents operated in the commercial
sector to discourage the construction of new shops and offices
in city centres. Davidson and Leonard (1974) stress the role
of the propriété commerciale, a form of commercial lease
introduced after the First World War with the purpose of extending to small shopkeepers and artisans a similar security of
tenure to that enjoyed by domestic tenants. Rigidity was
introduced into the property market by the fact that the lease
could be renewed for a period of years at the sole request of
the tenant. Leases were sold but at a high price which, when
combined with the cost of the property itself, imposed a heavy
financial burden on any company wishing to carry out a
redevelopment scheme. Change and renewal were therefore slow
to take place and inner city areas, especially, retained an
intricate mixture of land uses. The demand for land free

from restrictions also meant that there was a low ratio of open space to number of residents.

The difficulty of freeing land within the city for development meant that the construction which did take place between the wars tended to be on the urban fringe, often distant from services and amenities. This was particularly the case outside Paris where small plots - lotissements - were acquired, either of unused land or by severance of agricultural holdings, and an untidy scatter of cheap dwellings resulted (Chapter 8). Parallels are suggested with the shanty towns on the edge of cities in the Third World. There was some tendency for development to follow the main line railways but journeys to work were long, and in social terms there was a total absence of that vie du quartier that characterized the crowded inner districts of French towns and cities. Subsequent planning was also made difficult by the haphazard distribution of individual dwelling units.

The slow pace of interwar construction, coupled with wartime damage (450,000 dwellings destroyed), meant an acute housing shortage in the 1940s and grave problems for the many young families that were now seeking accommodation. Construction in general was boosted by the fact that the cement industry was one of the six sectors of the economy chosen for investment under the Monnet Plan, but the need to restore the infrastructure and to encourage manufacturing ensured that in the early postwar years the principal beneficiaries were roads, power stations and factories rather than housing. As late as 1952 the number of new dwellings completed in a year was only 75,000.

By that time, however, the foundations of the subsequent housing drive had been firmly laid. These rest on what Duclaud-Williams (1978) has called the three pillars of postwar housing policy, pieces of legislation enacted between 1947 and 1950 with the intention of freeing the industry from the worst of earlier restraints and of encouraging new building. The first of these, introduced in September, 1947, re-constituted the public housing authorities, the HLMs, with the purpose of stimulating the construction of low-cost housing for renting. This was followed, a year later, by legislation aimed at relaxing the strict controls on rents that could be charged by private landlords. Controls were not removed altogether, but from 1 January, 1949 it became possible to increase rents on the basis of regular review. Finally, encouragement was given to the owner-occupied sector when, in July, 1950, the government made arrangements to guarantee the loans made to house-buyers by the Crédit Foncier, the principal organization responsible for this kind of credit in France.

The effect of the legislation was not immediate but the framework had been established for the housing drive which

began in the mid-1950s. This followed the serious concern
that had been expressed over the state of French housing
revealed by the census of 1954. At that time more than 20%
of the population was classed as living in acutely overcrowded conditions and many dwellings lacked the usual amenities, even an internal supply of running water (over 40%).
Responding at last to the crisis, the construction industry
was completing 300,000 new dwellings a year by the end of the
1950s. This total rose to 400,000 by the mid-1960s and
reached half-a-million for the first time in 1972. A maximum
of 550,000 was achieved in 1975, since which date construction
has slowed in response to the recession, but the number of
dwelling units started annually was still around 400,000 in
1980.

Rented Housing

        The distinction between public and private housing is less
clear in France than it is in Britain and a meaningful division
of the French housing market must take into account, not only
tenure, but also the level of state subsidy accorded to
developers and the extent to which rents are subject to control.
        Parallels may be drawn between the role of the HLM agen-
cies in France and that of the council in British towns, particularly in respect of their 'social' obligation to provide
cheap housing, but in the former case the involvement of the
municipality tends to be less direct. Conversely the state
has a greater say in when and where public housing will be
built in France.
        The HLM movement has its origins in the Loi Siegfried of
1894 which recognized the need for a measure of public intervention in the provision of housing and made government money
available to philanthropic housing societies. Similar bodies,
such as the Peabody Trust, were set up in the United Kingdom
but their function was largely taken over by local authorities
after the First World War and it is only in recent years that
semi-private and charitable organizations have re-emerged in
the form of housing associations. But in France, where there
was nothing comparable with Britain's Housing Act of 1919, the
provision of low-cost housing has continued to be the responsibility of distinctive organizations. Of these the most
important are the organismes HLM, but 'social' housing is
also built by other bodies, including the mixed economy companies that feature prominently in many aspects of French
planning.
        The various HLM agencies, of which there are some eleven
hundred, have been responsible for about 30% of the housing
built in France since the Second World War and the HLMs now
account for between a quarter and a third of all housing in
urban areas of more than 10,000 population (Hans, 1974), a
total of about 2.8 millions. Construction is financed largely

90

on the basis of state loans which are made available at low
rates of interest (2-3%) and are repayable over a long period
(40 years or more).   But there are some funds also from the
private sector, often, in the case of companies, in return for
an allocation of housing.   The rents charged are, as the name
habitations à loyer modéré suggests, below those that would be
obtained on the open market and both central government and
local authorities play some part in determining them.   Inevi-
tably, however, rents become a subject of lively and continuing
political debate, not least because they must reflect high
land prices and rising maintenance costs as well as the actual
costs of building.   The state also exercises controls through
the imposition of minimum standards on construction and by
appointment of leading administrators to the boards of the HLM
agencies.
    The HLM organizations have attracted criticism over the
allocation of housing, in particular for not carrying out their
obligation to house families in greatest need.   Fear that poor
tenants will not be able to pay even a low rent is undoubtedly
one of the reasons why HLM housing has been made available to
the better-off as well as to low-income families, but political
considerations have probably also played a part (Ardagh, p.328).
Confusion over their role arises, too, from the variety of
bodies represented within the Union des Fédérations
d'Organismes HLM.   The 1,120 bodies that made up this
federation in 1978 belonged to four main groups:

offices publics	298
sociétés anonymes (limited cos.)	380
sociétés de crédit immobilier	180
sociétés coopératives	262

Some are much more dynamic and better funded than others and
there is little collaboration despite their supposedly common
aims and the guidance which they receive from the federation
and an annual congrès HLM.   They also differ in what they
consider their social obligations to be, some having far more
in common with private development companies than with low-
profit housing associations.
    Attempts were made by the government in 1958 to influence
the allocation of tenancies.   HLM housing was not to be offered
to tenants whose income was above a certain level, and existing
tenants could be required to pay a supplement on top of their
rent if their means rose.   But the legislation was largely
ineffective since the income ceiling chosen was unrealistically
high and most of the HLMs proved unwilling to charge the extra
rent.   Little was achieved, in fact, until 1975 when the whole
system by which cheap housing was financed was investigated by
a group headed by Raymond Barre, soon to become Prime Minister.
The report of the Barre Commission highlighted the injustices

in the allocation of HLM housing and recommended an entirely different approach to housing subsidies. Instead of giving financial support as hitherto mainly to the bodies which actually built the dwellings (aide à la pierre), it was proposed that funds should now be directed towards the tenants. Under such a scheme for personal assistance (l'aide personnalisée au logement) it was hoped that the lowest-income families would enjoy better access to 'social' housing. It was also felt that, by giving the individual greater freedom of choice, it would be possible to reduce urban social segregation. The idea was taken up by M.Galley, Minister of Equipment, whose Bill on housing finance was approved early in 1977, giving improved financial support to tenants and establishing a common system of rents amongst all HLMs. Unfortunately the legislation was introduced when the demand for this kind of housing appeared to be declining. The HLM agencies had completed 115,000 dwellings for rental in 1975 but by 1978 this total had fallen steadily to 69,000.

Much of the HLM housing built in the 1950s and 1960s took the form of monotonous high-rise blocks of flats, frequently grouped in grands ensembles. In official parlance the term 'grand ensemble' is applied to housing estates of at least 8,000-10,000 dwellings, housing some 30,000-40,000 people, but it is popularly used to describe groups of probably only a few hundred dwellings where housing consists exclusively of apartment blocks. This kind of construction was favoured in other countries too, of course, but the emphasis on the collectif was particularly strong in France. Here it may be attributed in part to the severity of the housing crisis and the attraction offered by industrialized methods of construction for meeting the need. But it also reflects the influence in architectural matters of what is referred to variously as the rationalist, progressive or modern school.

The buildings of the progressive school were simple and austere, based on pure, geometrical forms and erected in modern materials - steel, concrete, etc. In urban design there was a strict separation of land uses with each activity allocated its place according to a predetermined plan. The road, a symbol of disorder in the existing town, gave way to open spaces and a green setting. Unquestionably the most influential exponent of these ideas of the modern school was Charles-Edouard Jeanneret (1887-1965) better known as Le Corbusier, whose models satisfied a certain preference in France for the grand design. His unités d'habitation also appealed to social theorists, who saw these giant constructions as modern versions of Fourier's phalanstère, collective dwellings that would impose a new order on society, replacing individualism with a responsibility to the group, to the community. The first such unité, or ville radieuse, was built at Marseille between 1947 and 1952 and attracted much popular criticism as 'une sorte de centre

d'élevage humain' (Choay, 1965).   More influential, especially
amongst professional architects and planners, were Le Corbusier'
writings, including his <u>Manière de Penser l'Urbanisme</u> (1946)
in which he anticipated that 'la ville se transformera petit
à petit en un parc' (p.86).

The adoption of ideas derived from the progressive school
was made possible by the introduction of industrialized methods
of building.   Standardization of production favoured simple
architectural forms, the large box enclosing units of standard
dimensions that could be assembled rapidly from factory-built
panels and fittings.   It was the age of the module.   Industrialization of house-building also lowered costs, and Simonetti
(1978) gives the example of a 3-room HLM dwelling, the average
cost of which represented 54 months' wages in 1960 but the
equivalent only of 37 months' earnings ten years later.
Simonetti also associates the industrialization of the construction industry with what he describes as 'un urbanisme de
grand ensemble'.

Work began on the <u>grand ensemble</u> of Sarcelles to the north
of Paris in 1956.   Others quickly followed, in Paris and elsewhere, and the preference for multi-storey flats extended even
to quite small towns where one or two blocks would be built
incongruously on the edge of the urban area.   Where tower
blocks are grouped together in <u>grands ensembles</u> the layout is
usually one incorporating large areas of open space between
the individual buildings.   Yet despite the open nature of the
planned layouts, residential densities are often high, reflecting the small size of individual dwelling units with many
households having no more than three, possibly four, rooms.
In his study of housing and housing density in France, Rapoport
(1968) refers to the average occupancy rate for HLM housing of
1.15 persons per room, but finds rates as high as 1.38.   Such
high rates of occupancy mean that gross densities often exceed
100 persons per acre, and may reach 200 persons.   Rapoport is
highly critical of the layout of these estates, referring to
vast, dead, open spaces between huge slabs of buildings, and a
feeling of isolation to which the wide, green expanses give
rise (Figure 12).   Animation and urbanity, which one might
have expected to be associated with high residential (or
physical) densities, has been lost in the open spaces which
give low visual densities.   Rapoport attributes the confusion
between physical and visual density to the school of
Le Corbusier, describing the latter's plans as garden city,
neglectful of the French vernacular tradition, and anti-urban.
Other writers, however, have seen a less direct connection
between Le Corbusier's work and the design features of the
<u>grand ensemble</u>, preferring to attribute the failings of the
latter to a misinterpretation of Le Corbusier's ideas.

High land costs in French cities (below) ensured that
many of the <u>grands ensembles</u> were built on the urban fringe,

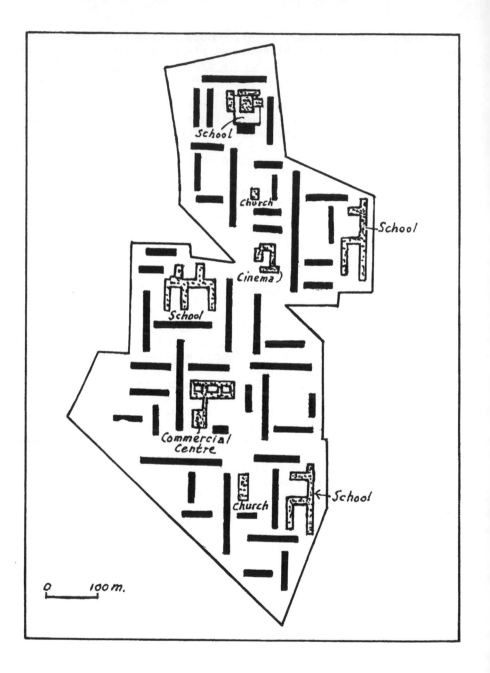

Figure 12. Grand ensemble of Poissy-Beauregard, after Rapoport, courtesy of Town Planning Review

94

often far from work and ill-served by public transport and
other services.   Commonly sited on the interfluves between
river valleys (to Le Corbusier the view helped one to identify
with nature), they stand out starkly against the skyline and
their geometrical outlines are now a familiar feature of the
approach to any sizable French town.   To their critics the
tower blocks are little more than 'silos à main-d'oeuvre', and
the individual flats 'cages à lapins' (rabbit hutches).   They
have also contributed to social segregation since it was the
policy of the larger HLM agencies to build different categories
of housing for different tenants (Duclaud-Williams, 1978).
For the families displaced by the clearance of shanty towns
(bidonvilles) and for those from the worst of the urban slums
there was a special 'heavy' kind of dwelling.   Lower rents
were charged for these flats, but they were often in blocks
isolated from other types of HLM housing.

     In the 1960s the grands ensembles had a fairly typical
'new town' population which included large numbers of young
married couples and their small children (Clerc, 1967).   They
also housed pieds noirs returning from Algeria after 1962.
Despite the well-publicized problems of lonely young wives and
inadequate facilities for growing children, the shortage of
housing ensured long waiting lists at the offices of the HLM.
Most of the blocks were of three to seven storeys, but some
were higher.   They were hastily constructed, however, and
time has revealed a wide range of structural defects.   The
result has been an increasing disenchantment on the part of
middle class families who were attracted in the early years
by new housing at relatively low rents.   As they have moved
out, they have been replaced by immigrant families and other
under-privileged groups.   'Nous avons entassé dans ces
ghettos les sous-prolétaires du XXIe siecle' (Hubert Dubedout
in L'Express, 26 February, 1982).   Crime and other social
problems have increased, leading, in the worst cases, to the
abandonment of whole blocks.

     Growing public disquiet about the tower blocks brought
the first official reaction when, in November, 1971,
M.Chalandon (Ministre de l'Equipement et du Logement) produced
a government circular forbidding the construction of such
buildings in towns of less than 50,000 population.   Rules
were also drawn up which sought to control the size of blocks
elsewhere and to ensure a better provision of lifts and communal
facilities.   An extension of these improvements was proposed
when, in March, 1977, an inter-ministerial group, 'Habitat et
Vie Sociale', was set up.   The Seventh National Plan included a programme of works to be carried out by the group in
selected urban areas under the general title of 'Mieux Vivre
dans la Ville'.

     Government funds have also been made available to repair
structural deficiencies where this is possible.   Despite these,

known as <u>primes à l'amélioration des logements à usage locatif</u>
<u>et à occupation sociale</u> (PALULOS), it was reported in 1981 that
out of a total of 2.8 millions, 300,000 dwellings were 'en
danger' and a further 300,000 were approaching a 'situation
critique' (<u>Le Monde</u>, 3 November, 1981).   So great was the concern felt at this situation that in December, 1981 the new
socialist government established a Commission Nationale pour
le Développement Social des Quartiers, under the presidency of
M.Dubedout, mayor of Grenoble, in order to carry out an emergency programme of rehabilitation.   Sixteen sites were designated as priority areas for action during the first year and
the first contract, permitting work to begin, was signed in
February, 1982.

The list of 16 includes some of the <u>grands ensembles</u> that
have become notorious over the years for their problems: Les
Minguettes at Vénissieux outside Lyon, 'Les 4,000' at La
Courneuve in the suburbs of Paris, Le Haut-du-Lièvre at Nancy,
400 metres long and popularly known as the Great Wall of China,
and Le Quartier Paul-Mistral at Grenoble.   Les Minguettes,
isolated on its plateau site, more than an hour's bus ride from
the centre of Lyon, has a population totalling 30,000.   Here
the greatest problems centre on a district known as Monmousseau
where of nine tower blocks built as recently as 1967, three
have been wholly abandoned, and the incidence of vandalism and
crime has attracted the attention of the national press.   La
Courneuve in the <u>département</u> of Seine-Saint-Denis was built
on a 37-hectare site between 1963 and 1968 and its 4,000
dwellings have given it the popular name of 'Les 4,000'.
Many of these are in 15-storey tower blocks and its population
of 17,000 represents 43% of that of the commune in which it is
situated.   Thirty per cent are immigrants and nearly half of
the total are under the age of twenty.

In some cases the programme of works involves demolition,
but the intention where possible is to rehabilitate existing
dwellings.   In addition to carrying out physical repairs, it
is hoped also to improve the provision of public transport,
sports and other facilities, and encouragement is being given
to organizations engaged in community work with a view to promoting better relations with the police and between immigrants
and older residents.   Financial measures are also envisaged,
increasing the funds available to the HLM agencies and
extending assistance with rents (the APL) to those in greatest
need.

The above account has concentrated on the 'public' or
'social' sector of the French housing market, much, though by
no means all of which is to be found in the postwar apartment
complexes or <u>grands ensembles</u>.   Outside this category,
property is also rented from private landlords and from a
variety of institutions, including banks and insurance companies.
The level of rents ranges widely within this 'private' sector,

depending to a large extent on whether the housing concerned
was built before or after 1948.   As a result of the 1948 Act,
a relatively free market was introduced in new property,
whilst controls remained, albeit of a less rigid kind, on
older dwellings.   The effect was to bring about great contrasts in the amount of rent charged, differences that did not
by any means always reflect the quality of the accommodation
concerned.   In order to deal with the more obvious anomalies,
the government has tried, through successive pieces of legislation, to de-control selected categories of property, for
example dwellings in towns of under 10,000 population (1958).
As a consequence of these measures the number of properties
subject to the controls imposed in 1948 has fallen from 6
million soon after the Act came into force, to 3.3 million in
1966 and 1.1 million in 1975 (Duclaud-Williams, 1978).   The
highest proportions of rented dwellings subject to control are
to be found in the inner parts of the major cities, especially
in Paris.

There is government support for the private rental sector
which mainly takes the form of grants to carry out repairs.
These are paid by the Agence Nationale pour l'Amélioration de
l'Habitat (ANAH).   With the aim of adding to the overall
stock of housing, a number of financial concessions are also
available to companies building property for letting.

The Owner-occupied Sector

For the individual buyer there is nothing comparable in
France with the British building society movement.   Loans for
house purchase are obtained from banks and savings schemes
(e.g. Crédit Agricole) and from a body known as Pret Immobilier
Conventionée, but by far the most important source of finance
for this purpose is the Crédit Foncier, a mortgage bank which
is organized on the same basis as a private institution but
which operates under guarantees from the government and has
its chief officials appointed by the state.   Between a
quarter and a third of the dwellings erected by HLMs are also
sold to owner-occupiers and a system of loans is operated by
these agencies for the purpose.

In order to assist house purchase and to stimulate construction, the government has made it possible for the borrower
to obtain financial assistance with interest payments.   This
takes the form of a subsidy which is related to the floor space
of the dwelling.   For his part, the builder who borrows money
to finance a development is also able to obtain loan guarantees
from the Crédit Foncier even if the funds have been secured
on the private market.   In return for this government assistance, the developer is required to limit his profit margin to
an agreed percentage, the overall aim being to exercise some
control over the price of new housing.

Approximately half the dwellings in France are now owner-occupied, the proportion having grown from a third in the mid-1950s.   The trend towards owner-occupation has been particularly marked in the inflationary 1970s and this is reflected in a marked shift in the construction industry away from blocks of flats towards individual houses.   Some apartments are built for purchase, especially at the luxury end of the market, but recent years have seen the re-emergence of a strong preference for the small house standing on its own plot.   In 1970, when 180,000 such houses were built, the proportion of flats to houses was of the order of two to one.   By 1980, when the number of individual houses completed had risen to 275,000, the ratio had been reversed, with two such houses to each flat or apartment.

The modern Frenchman's desire for a house surrounded by its own garden runs counter to a strong intellectual tradition in France which sees the pavillon as wasteful of land and energy and contrary to the 'Latin' practice of town-building. The opposition is partly aesthetic, offended by the untidy development of inter-war lotissements, and inclined to dismiss those who prefer a suburban villa as 'rurbains'.   But it also rests on a combination of political and social beliefs which represent the pavillon as an expression of individualism and contrary to the collective ideals of a perfect society (Osborn, 1967).   To the extreme political left the owner-occupied house is a tool of the employer, undermining the power of the proletariat by reducing their mobility and putting them in debt to financial institutions.   Examples are quoted like those of Longwy and Denain, both towns being victims of the iron and steel closures at the end of the 1970s.   In Longwy, where only 30% of the dwellings are owner-occupied, the resistance was greater than at Denain where no fewer than 80% of the properties are individual houses, many of them bought at the encouragement of the company (Usinor), and where there was a greater willingness to accept redundancy terms.

It is amongst the lower middle classes, enriched by the expansion of the French economy before 1973 but frustrated in their housing desires by the grands ensembles, that the preference for the individual (and individualized) house is strongest.   Wealthier groups, with better access to second homes, show a counter tendency to return to central city living following the rehabilitation of historic quarters, but this trend is small by comparison, and the demand for pavillons remains high.

There are two large groups in the French construction business (Groupe Maison Familiale and Phénix); otherwise the industry is characterized by a large number of small-scale contractors.   The trend to individual dwellings has been encouraged by the aid given to these smaller companies under the so-called 'Concours Chalandon' (Concours de la Maison Individuelle) of 1969, and by the assistance with house pur-

chase accorded to families following the Barre recommendations
of 1975. Official approval of this politique des pavillonnaires followed with the publication in 1978 of the Mayoux
report on 'L'Habitat Individuel Péri-urbain'. It acknowledged the strong preference on the part of most people for
a separate house ('chalandonnette') and called for realistic
government policies in favour of this kind of dwelling. An
example of such policies was the Act of 31 December, 1976 concerning the lotissement, envisaged as a more appropriate tool
than the ZAC (below) for planning the layout of individual
houses.

The Mayoux report followed the publication in 1977 of the
results of an enquiry into violence, criminality and delin-
quency conducted under the chairmanship of M.Alain Peyrefitte.
It had criticized high-rise buildings as contributing to the
'imbalances of urban life' and had recommended the building of
individual houses in satellite townships. Further support
came from a study of the new suburbs which observed that
pavillons tended no longer to be scattered indiscriminately
across the countryside but were now more commonly grouped in
small estates and in 'nouveaux villages' of around 50 houses,
sometimes more (Siran, 1978). Given this measure of approval,
it is scarcely surprising that the proportion of houses to flats
has increased steadily since 1976 (Figure 13).

Like the grands ensembles, the nouveaux villages of
individual houses have become a common feature of the urban
fringe of most French cities. Like the former they also
exhibit strong social contrasts, in particular between what
Barrère and Cassou-Mounat (1980) call 'villages à l'américaine'
(often referred to as 'hameaux' = hamlets) where the professional and managerial classes are strongly represented, and
the 'villages sociaux' which are working or lower middle class,
the refuges of those able to flee the grands ensembles. Like
the estates of tower blocks, they often fail to achieve a full
integration into the infrastructural fabric of the parent city,
many indeed being in the territory of peri-urban communes.
According to the Mayoux report a fifth of the urban population
of France already lived outside the central city commune by
1975.

But suburbs are also to be found within the city boundary,
depending on its geographical extent, and a more dramatic
indication of the extent to which these postwar housing developments have transformed the face of French cities is to be found
in a study for the Conseil Economique et Social, published in
1981 under the title of 'S.O.S. Banlieues'. This revealed
that 18 million people, one in three of the total French population, could now be classed as suburban.

Housing Estates

In 1958 the Minister of Construction, M.Pierre Sudreau,

Figure 13. Styles in 'pavillon' housing

introduced the zone à urbaniser en priorité (ZUP - priority
urban development area) in order to speed up house construction. He was concerned at the slow pace of new building
and saw this as a consequence, partly of the shortage of local
authority finance, and partly of the difficulty of assembling
sufficient land for a major housing development in a country
where respect for the private ownership of property is strong
and there is limited municipal ownership of land. Special
powers were needed that would enable authorities to acquire
land and pay for the infrastructure required before a major
housing development could go ahead. The Minister also felt
that it would be necessary to concentrate the effort in selected areas in order to gain maximum benefit from investment in
services and equipment and to avoid untidy sprawl of the kind
often associated with the more speculative type of development. To achieve this, developments exceeding 100 dwellings
would be forbidden outside a ZUP.

The zone à urbaniser en priorité was an attempt at
comprehensive urban development directed by the local
authority. Designation of a ZUP had to be approved by central
government which provided financial support needed for the
installation of the major services, but within the zone the
municipality was required to establish a plan and allocate land
uses. In order to facilitate this, the public authorities
were given the right of first refusal on land offered for sale
in the ZUP for four years from its designation (Darin-Drabkin,
1977). Land could also be expropriated, owners being compen-
sated on the basis of an average of market prices prevailing
over the previous five years.

Within a ZUP the actual development could be carried out
by the local authority but has generally been the responsibility
of a mixed economy company in which the municipality is represented, together with private capital, the involvement of the
latter usually being restricted to 50% of the total capital
requirement. This solution is intended to give flexibility
as well as attracting greater financial backing.

One hundred and ninety seven ZUPs were created between
1959 and 1969. After 1969 the ZUP procedure was largely
replaced by that of the zone d'aménagement concerté (ZAC),
although the right to designate a ZUP was not finally abolished
until 1976. Most ZUPs are on or near the edge of cities and
they have incorporated many of the grands ensembles favoured
during this period. When the legislation was introduced it
was envisaged that a ZUP would include land sufficient for at
least 500 dwellings. Many, in fact, are much larger than this.
In about 50 of them the number of dwelling units built exceeds
5,000 (Barrère and Cassou-Mounat, 1980), and the average is
between 2,500 and 3,000, a total of almost 570,000 dwellings
having been erected within ZUPs between 1959 and 1975. The
largest is that of Toulouse-Le Mirail with 23,000 dwellings

on its 800-hectare site; other large ones with populations ranging from 25,000 to 40,000 include Les Minguettes outside Lyon, Grenoble-Echirolles, and Herouville Saint-Clair at Caen.

The ZUP was a response to a pressing housing need. Individual estates were built mainly of apartment blocks and with a high proportion of HLM housing. They were often far from places of work, and although many of them were the size of small towns the provision of services was slow, attracting much adverse comment. A typical example is the ZUP of Bellevue at Brest, begun in 1963 by the Société d'Economie Mixte pour l'Aménagement et l'Equipement de la Bretagne (SEMAEB) and already housing more than 20,000 people by 1974. But complaints about services mounted and in that year the municipal council of Brest was forced to announce a programme of works that included better shops, health care and schools, together with the provision of recreational open space, car parking and road signs ('Dans la ZUP, on y entre mais on ne sait pas comment en sortir' - Jean de Rosières, Le Monde, 22 April, 1974).

Particularly harsh criticism of the ZUPs was voiced by M.Guichard in a ministerial circular published in March, 1973. He saw them isolated from the life of the rest of the city, the resources of which they had helped to drain: 'la ZUP de taille si disproportionnée à l'agglomération mère qu'après en avoir absorbé toute l'énergie et les ressources elle se traîne comme un corps inerte à son flanc, incapable de s'intégrer à sa vie comme d'en acquérir une propre'. They had increased social segregation by their emphasis on HLM housing, and M.Guichard blamed local authorities who saw electoral advantage in concentrating votes in certain wards of the city: 'les municipalités qui entassent les logements HLM comme un petit trésor électoral font, malgré de fortes apparences et de vigoureuses déclarations, le contraire d'une politique sociale. C'est une politique de sécession, alors que nous visons à l'intégration'. It was comment of this kind which, a few years earlier, had led to the introduction of the ZAC in place of the ZUPs and their gigantisme.

Another criticism of the ZUP was that it fuelled speculation in land just outside its boundaries, property developers hoping to take advantage of the potential market afforded by the nearby population and of roads and other equipment built to serve the estate. It was to overcome this problem, and also to avoid the necessity of local authorities having to purchase large amounts of land for development all at once, that the zone d'aménagement différé (ZAD - zone of deferred development) was introduced in 1962.

The ZAD is an area within which it is proposed to carry out some form of planned urban growth, its boundary being more extensive than that intended for actual development in order to avoid the problem of speculation around the edges referred to above. If licences for building already exist within the defined zone these cannot be withdrawn, but otherwise all rights to develop are frozen in anticipation of a set of major proposals for public development, usually a ZUP or ZAC. For a period of 14 years (it was originally 8) the local authority has pre-emption rights on land, i.e. the right to acquire land that is offered for sale and, most importantly, the right to purchase such land at a price equal to its value one year before the intention to declare the ZAD was made public. The idea behind this clause is to avoid the kind of speculation that took place when proposals for a ZUP were leaked and so protect the local authority from the necessity of paying exorbitant sums in compensation. In practice the purchase price is usually a matter of negotiation between the public authority and the existing owner. A sum in excess of the 'frozen' price may be paid but the difference is not great. Powers of compulsory purchase may also be exercised when required and this is done by making a déclaration d'utilité publique.

By 1977 over half-a-million hectares of land had been zadés. In a few rare instances, as at Rennes, a local authority has tried to control urban sprawl by subjecting the whole of the urban fringe to this procedure. But the cost involved has generally made such a practice prohibitively expensive.

A modification was made in 1965 to the rules relating to compensation, the intention being to benefit the private land-owner whose expectation of future profit on land had been denied him. Now an owner whose land had not been purchased after a period of three years from the declaration of the ZAD could ask for it to be revalued and this revaluation was per-mitted to take account of the normal rise in land prices over the period. It was a concession to the private sector that offended those who wished to see the land market brought under stronger public control and who pointed to the difficulty already experienced by local authorities in finding sufficient funds for their land requirements.

Similar differences of opinion have been expressed about the zone d'aménagement concerté (ZAC - zone of concentrated development), introduced in 1967, which soon replaced the ZUP as a basis for carrying out large-scale urban developments. Cooperation between public and private interests is central to the planning of a ZAC. Supporters of private enterprise see this as a useful means by which national or local govern-ment can draw on finance from the private sector, whilst those committed to public ownership are inclined to condemn it as another example of the power of capitalism to exploit the public interest.

The ZAC formula has been used to carry out schemes of commercial redevelopment in city centres and in association with the building of industrial estates, tourist resorts and university campuses. But it is most commonly employed in connection with housing projects as an alternative to the earlier and less flexible system of the ZUP. The public interest is usually represented by a local authority, but in the larger projects this may be the state or some form of public or semi-public development corporation. The degree of public involvement also varies. In some cases the local authority acquires the land and makes itself responsible for all the infrastructure and services, including shops and schools, leaving only the housing to private developers. But in others the land remains in private ownership and the promoters make a contribution towards the cost of major services in return for the right to develop and, probably also, certain tax concessions. The nature of this contribution is laid down in the contractual agreement drawn up with the local authority and may include a financial payment or the gift of land or buildings to the authority for public use. The whole is covered in a development plan, this and the contract (convention de ZAC) having to be approved by central government.

It was hoped by means of the contractual agreement to avoid many of the problems that had arisen in the case of the ZUPs. The respective contributions of the different bodies, public and private, would be more clearly defined and it was expected that there would be none of the delay in the provision of services that had caused such difficulties on the earlier housing estates. Another difference between ZUP and ZAC lay in their size. The former, it was generally agreed, were often too big in relation to their parent city, and although some ZACs may include as many as 5,000 dwellings, the average is much less, nearer one thousand. Indeed, since March, 1973, the size of ZAC has been restricted by law to no more than 1,000 dwellings in towns of which the population does not exceed 50,000. The limit is 2,000 housing units in larger towns. The same legislation sought to reduce the degree of social segregation that had characterized the ZUPs by requiring that a ZAC estate should have at least 20% of its housing built by HLM agencies for letting but that this proportion should not exceed 50% of the total. It was also suggested, with a view to introducing variety in design and appearance, that no one architect should be responsible for more than 500 dwellings on an estate. Earlier, in 1971, M.Chalandon had issued a ministerial circular which decreed that at least 50% of the dwellings on a ZAC estate in a small town (under 20,000 population) should take the form of individual houses, and that there should be at least 30% in towns of under 50,000. The intention was to replace 'l'urbanisme sauvage' of the ZUP with 'un urbanisme à l'échelle humaine'. Finally in the attempt to achieve this, it was laid down in 1974 that no ZAC would be permitted where there

104

was not an agreed town plan (SDAU - below) and that no ZAC could be built outside the area covered by such a plan.   More than 1,600 ZACs had been defined by the end of 1977, accounting between them for some 83,000 hectares (Claval, 1981).

Land Use Planning

The priority zone, represented by the ZUP and more recently the ZAC, has been widely employed as a tool for carrying out large-scale urban development projects.   But the location and, to a certain extent the layout, of such schemes needs to take into account those plans that have been drawn up for the urban area as a whole.   In France the Town Planning Act of 1943 anticipated the necessity of planning for postwar reconstruction by introducing the plan d'urbanisme.   That legislation was reinforced in 1954 and again in 1958 when different categories of plan were distinguished, in particular the plan directeur and the plan de détail.   Plans drawn up on the basis of this 1958 legislation have influenced the development of French towns, in some cases until recent years, but their failure to establish strong and lasting guidelines has already been noted in the previous chapter (Comby, 1975).   At the very time when major new problems were being created by rapid urban growth, plans were often abandoned and new ones delayed by conflicts of interest over land and other matters.   By the mid-1960s there was urgent need of a stronger and more comprehensive system of town planning to enable local authorities to control the physical development of cities.   Under M.Pisani as Minister of Equipment, the Loi d'Orientation Foncière et Urbaine was approved in December, 1967 in order to establish this degree of control.   It did so by introducing the schéma directeur d'aménagement et d'urbanisme (SDAU) and the plan d'occupation des sols (POS).

The SDAU is broadly comparable with a British structure plan in that its intention is to lay down broad guidelines rather than to establish detailed land uses.   It may cover a single commune or a group of communes and its purpose is to indicate for both the medium and long term (20-30 years ahead) which areas should be developed and for what purpose and which should be protected.   These recommendations are set out in a written report and maps, and there is an obligation to produce a SDAU on the part of all agglomerations with a population in excess of 10,000.

In the preparation of a schéma directeur, the local authorities receive advice from interested public bodies and strong directives also from central government in the person of the departmental or regional préfet.   The latter must ensure that the blueprint is in line with the forecasts made in the national 5-year plans and takes into account any major projects such as motorways or regional parks which are being proposed by the government or the regions.   When it has been completed the SDAU must, however, be approved by the municipal councils of the communes concerned. Plan-making is a time-consuming process and of 412 SDAU on which work had been undertaken, only 131 had been finally approved by October, 1978 (Pagès, 1980).

The plan d'occupation des sols is intended to complement the SDAU in much the same way as a local plan complements the structure plan in Britain. It is a detailed land use plan, defined by the Ministry of Equipment (in a decree of 4 November, 1970) as 'un document de synthèse destiné à régler l'ensemble des problèmes relatifs à l'utilisation des sols'. The plan lays down precisely the limits of those areas within which it is permitted to build and those which must be kept free of construction by reason of their value to agriculture and mining or because of the quality of their scenery. Areas subject to hazards such as flooding or avalanches are also demarcated.

Amongst the areas classed on the plan as 'zones urbaines' a distinction is made between different types of construction considered permissible, for example between 'construction en ordre continu' and 'construction en ordre discontinu'. To maintain this distinction an important part of the plan is the coefficient d'occupation des sols(COS) which is applied to the individual districts or zones of the plan and fixes the maximum permissible density of development. The COS is based on the relationship between floor space and total ground area. Thus a COS of 3 would allow building of 3,000 sq.m of usable space on a plot of 1,000 sq.m in area. Financial penalties are incurred if the permitted densities are exceeded and the building coefficient is applied to sites both in public and in private ownership.

The POS, once approved, is a document that is legally binding on both the individual owner and the local authority. Like the SDAU, to the general guidelines of which it should conform, it can apply to a single commune or more than one, and it is important in controlling the future use of land, licences for building being granted in accordance with this plan. Under these circumstances the plan cannot establish long-term rules and the POS is therefore a relatively short-term document, covering a period of five to ten years. In order to provide flexibility and to encourage the housing drive, the 1967 Act also laid down that areas designated as ZACs should be exempt from the building regulations applicable in a POS. Despite this, the POS has been criticized for its rigidity in areas where growth pressures are strong.

The system of schéma directeur and plan des sols introduced after 1967 has attracted criticism on other grounds too. It has been described as quantitative rather than qualitative, typical of the 1960s when the need to find land suitable for housing and other forms of development was stronger than the

desire to protect the environment.   Significantly 1966 saw
the creation of the Ministry of Equipment, giving a prominent
role in the planning process to engineers who were inclined to
look upon the urban fringe as a reserve of building land rather
than some kind of green belt affording protection to the growing
town (Barbier, 1978).

The length of time required to draw up the plans has also
attracted adverse comment.   It was intended in 1967 that all
plans should be completed by 1 January, 1975, but this proved
to be far too optimistic and the deadline has been successively
extended.   Pagès (1980) notes that although work had been
undertaken on 9,248 POS, only 3,097 had been made public by
October, 1978 and only 425 finally approved.   In these circumstances it is hardly surprising that the plans were often
overtaken by events and were, in several respects, outdated
before they were completed.   Since 1978 considerable efforts
have been made to speed up the completion of the POS and 6,800
had been published or approved by the end of 1982, but snags
are still encountered especially in smaller communes.

Amongst the principal causes of delay have been the
rigidity of local government structures and the strength of
land and property interests.   The first of these relates to
the very large number of communes and the difficulty of
achieving cooperation despite legislation aimed at encouraging
the fusion of the smaller ones and at the creation for planning
purposes of <u>districts urbains,</u> <u>communautés urbaines</u> and <u>syn-
dicats intercommunaux</u>.   Friction between communes, even within
these wider groupings, is common as we have noted in earlier
chapters, and the cherished independence of the commune remains
a powerful force despite attempts to set up new inter-communal
organizations.

The land question has probably been the greatest stumbling
block to urban planning in postwar France.   It is seen not
only in the complicated pattern of ownership but also in the
high cost of building land and in the steep rise in land
prices as a result of speculation.   Pagès (p.37) refers to an
18-fold increase in land prices in Paris between 1952 and 1969
and a situation in the Paris Region where no less than 40% of
the cost of construction is accounted for by the cost of land.
Outside Paris the corresponding increase was 8-fold, with land
accounting for 20% of total costs.   Various attempts were made
to curb the speculation, including the introduction of the
<u>zone d'aménagement différé</u> in 1962, and a plan (1971) to dis-
courage owners from holding on to unused plots in city centres.
But the overall success of these measures was limited and the
first comprehensive strategy for controlling soaring land
values was the Loi Foncière (Urban Land Bill) of December, 1975
which became operational in April, 1976.

The Loi Galley, which takes its name from M.Robert Galley,
Minister of Equipment, who introduced the legislation, was an

attempt to extend local authority control over urban development by limiting the power of land and property speculators. It sought to do this by imposing a limit on the permitted density of development and by the introduction of the zone d'intervention foncière (ZIF).

For Paris the legal limit on density (PLD = le plafond légal de densité) was fixed at 1.5. Elsewhere it was to be no higher than 1.0, meaning that on any plot it was permitted to build an amount of floor space equivalent only to the area of that plot. The legal limit would be reached, for example, if on a plot of 1,000 sq.m a 5-storey block were to be erected with floors of 200 sq.m. Developers who exceed the PLD are required to pay a financial penalty. This tax is equal to the purchase price of the amount of land by which the development is in excess of the permitted limit, three-quarters of it being paid to the local authority concerned and the remainder to a national local authority fund set up to ensure that some of the profits of land speculation will be more widely shared amongst the communes.

M.Galley described his legislation as 'une loi pour l'environnement', claiming that, 'le temps du béton à n'importe quel prix est aujourd'hui révolu' and 'moins de rénovation sauvage qui brise l'harmonie de la ville, moins de ségrégation entre les riches et les pauvres' (Le Monde, 24 October, 1975). One of his main concerns was with the effects of high-rise office development on city centres. Visually intrusive, such office towers were attracting growing hostility, especially in Paris, and they were also the principal cause of speculation in land. Of the construction that was taking place in France in the early 1970s, only one-tenth of the new floorspace was in city centres but this accounted for two-thirds of all the money spent on land for building.

A related concern was with housing and social segregation. The 1975 census revealed falling population totals in central cities and it was argued that the price of land was driving the better-off to the new low-density estates in the suburbs, leaving the once-fashionable central areas increasingly to the elderly, the very poor and the immigrants. Critics of M.Galley's loi were inclined to lay stress on the supply of land, however, rather than its cost, estimating that the PLD would reduce still further the supply of building land and thus encourage gentrification by ensuring that only the luxury housing market was catered for in new developments. But in either case the effect was likely to be increasing social polarization within the central cities.

One must be wary of supposing that there is a simple relationship between the movements of population on the one hand and, on the other, such factors as the supply of building land, its cost, and the rents charged for housing. But whether they live in crumbling property in the centre or in tower blocks on the periphery, it is likely that the poor will suffer most from speculation and that the local authority will be hampered in its efforts to bring about urban improvements by the difficulty and cost of acquiring land. The zone d'intervention foncière stemmed from an awareness of these problems, its aim being to enable local authorities to carry out 'social' policies involving new housing, the restoration of old quarters, and the provision of open space, by acquiring both land and the profits from land deals for the benefit of the community. The ZIF has much in common with the ZAD but is typically used where some form of redevelopment is proposed, rather than on virgin sites. Within a ZIF the local authority possesses the right to purchase land, either at the market price, or at a price that can be arranged by arbitration, and the latter may be based on prices current the previous year. There is no time limit on the power to acquire land in this way. It was expected in 1975 that a part, at least, of the funds needed by local authorities to purchase land would come from the application of the PLD, this section of the Act thus being seen as complementary to that of the intervention zone. In fact, such funds have been limited, office developers in the recession years preferring to keep close to the permitted densities.

More than 600 ZIFs had been declared within two years of the adoption of the Loi Galley and the number has grown since. This is partly due to the fact that a ZIF applies automatically to all those areas which are classed as 'zones urbaines' on an approved or published POS. ZIFs can also be created in small towns at the instigation of the préfet.

In common with other countries in the West, the French government has attempted by the introduction of successive measures since the Second World War to curb speculation in urban land. These have been interventionist measures, stopping well short of the nationalization of building land that some may have wished for on the far political left. Furthermore it is unlikely that the election of a socialist administration will move much further in this direction in a country where the right to profit from one's lopin de terre is such a deeply held belief. The Loi Galley has enabled local authorities to exercise a degree of control over the urban land market, assisting plan-making, but the measures were too late to check the soaring prices that gave to French cities in the 1960s and early 1970s some of the characteristics of urban settlements in the Third World. The legacy of these years is to be found in the speculative developments that mark the urban fringe and in the problems of renovation that beset the historic urban cores.

Chapter 6.

REVOLUTION IN RETAILING

There is no better illustration of the change that has
overtaken French cities in the last thirty years than the
transformation that has taken place in urban retailing.
Until as late as the mid-1950s the distributive trades were
still mainly in the hands of the small independent retailer or
the owners of long-established department stores. The former's interests were jealously guarded by right-wing
politicians such as Pierre Poujade, and the épicerie du coin
seemed as lasting a feature of French life as the small peasant
farm. But change, when it came, was rapid and within twenty
years the hypermarket, furniture store and garden centre had
become as typical of the approaches to most French cities as
the water tower, garage or apartment block. There were a few
examples also of the major regional centres that have become
so common in the urban fringes of American cities. Reaction
followed, however, and by the early 1970s increasing disquiet
was being expressed at the effect which the new stores were
having on established traders, especially in city centres.
Their scale, and the fact that the hypermarkets catered almost
exclusively for the car-borne shopper, were sources of concern
in a more environmentally-concious age. Legislation followed
to control their spread, and the emphasis is now on the small
neighbourhood centre and on renewal within the urban core.
    Before examining these changes in detail it is worth
recalling that the French were pioneers in the development,
more than a century ago, of another form of urban retailing,
the department store. The earliest of the famous Parisian
stores, Le Bon Marché, opened in the Rue de Sèvres in 1852, .
and this was quickly followed by Le Louvre in 1855 and Le Bazar
de l'Hotel de Ville a year later. Then came Le Printemps in
1865 and La Samaritaine in 1870. The effect of these stores
on their less competitive neighbours, and the way in which
they expanded to occupy the whole of island sites adjoining
the fashionable new roads, has been described by Zola in his
novel, Au Bonheur des Dames.

From Paris as centre of innovation, the department store
or grand magasin, spread to the larger cities throughout France.
Much later, in the inter-war years, the companies which ran
these stores began to open what have been called, somewhat
ambiguously, 'petits' grands magasins, or magasins populaires,
in the centres of smaller provincial towns. Les Galeries
Lafayette were responsible for the Monoprix chain, for example,
and La Société Printemps for the stores known as Prisunic.
These popular stores, of which there are more than 700 in
France, sell a wide range of consumer goods but at the cheaper
end of the range and with more emphasis on food than in the
case of the major department stores (Metton, 1980). Like the
latter, they have tended to remain loyal to city centre
trading.

Department stores range widely in size from the small
ones with about 1,000 sq.m of sales area to the biggest with
more than 10,000 sq.m of selling space arranged on several
floors. The Bordeaux store of Les Nouvelles Galeries offers
11,225 sq.m of floorspace, for example, and in Metz the same
firm has a store of 11,550 sq.m. The magasin populaire
rarely exceeds two floors and its selling area is typically
between 1,500 sq.m and 2,500 sq.m (Poittier, 1974). Both
grands magasins and magasins populaires continued to be built
in the years following World War II. In her study of grandes
surfaces in Lorraine, Poittier discovered that nine of the 15
department stores had been opened since 1945 and 19 of the 26
magasins populaires. Significantly, however, the last of
these popular stores opened in 1967, just as work was beginning on the first of the hypermarkets. Lorraine's first
hypermarket opened its doors in 1969; by 1973 there were ten
operating in the region.

Advent of the Hypermarket
The opening, in June, 1963, of France's first hypermarket
in the Paris suburb of Sainte-Geneviève-des-Bois heralded the
start of a revolution in the French retailing industry as far-
reaching in its impact as the arrival of the grand magasin in
the 1850s. By the end of 1981 there were 466 hypermarkets
operating in France, together with 4,335 smaller supermarkets.
Between them they had a sales area of some 6 million sq.m and
employed a staff of more than 200,000 (Libre-service Actualités,
nos.825 and 826, 11 and 18 December, 1981). Some of the new
stores have been built in town centres and the older suburbs,
but the great majority are located on the urban fringe, a
favoured location for the larger hypermarkets being close to
major road intersections, especially where routes nationales
and autoroutes approach newly-built outer ringroads (Figure
14).

The rapid adoption of this new form of trading can be
attributed to a range of circumstances which created a social

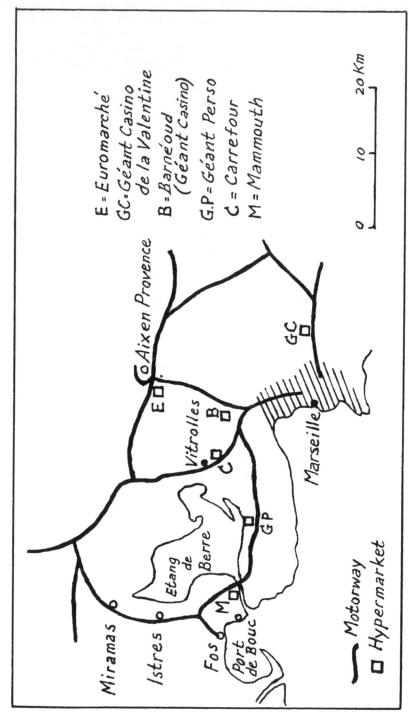

Figure 14. Location of hypermarkets in the Marseille area, after Vaudour, 1978

and economic climate favourable to the <u>grandes surfaces</u>
(Smith, 1973).   These have included:

(i)   growth and urbanization of the population, with its
        large new youthful element;
(ii)   increase in disposable income;
(iii)   growth in car ownership;
        The number of cars on the roads of France rose from
        about one million in 1950 to 18.5 million in 1980,
        equivalent to one car for every three persons in
        the country.   A growing proportion of families have
        two cars, the second almost certainly being one of
        those small cars that is favoured by the French road
        tax system.   The convenience of these vehicles for
        shopping is evident in the way they are advertised.
        Thus, 'elle supermarche bien', or the car is com-
        pared with a supermarket trolley and described as
        'une voiture-caddie'.
(iv)   change in shopping habits with a shift to bulk-
        purchasing, in turn due to such factors as the growth
        in the number of women in employment and the intro-
        duction of domestic storage appliances, including deep-
        freezers;
(v)   congestion in city centres and the availability of
        cheaper land on the urban fringe;
        Vaudour (1974) quotes the example of 'Euromarché-
        les-Milles', built on an unattractive site outside
        Aix-en-Provence where the cost of land was fifty
        times less than in the city centre.
(vi)   the earlier neglect of shopping provision in the
        peripheral estates, especially in the <u>grands ensembles</u>;
        Government concern for this under-provision of
        retail services was expressed in the 'Fontanet
        Circular' of August, 1961 which sought to persuade
        developers to include suitable shopping facilities
        in their plans (Beaujeu-Garnier and Bouveret-Gauer,
        1979).
(vii)   the slowness of French companies to adopt the kind of
        supermarket trading that characterized the suburbs of
        British cities in the 1950s;
        Smith (p.305) goes so far as to say that France
        missed a whole evolutionary stage in retailing.
        The effect was to add to demand when hypermarkets
        were introduced a decade later.
(viii) the inadequacy of planning controls which gave
        developers a much greater freedom over the choice of
        sites than retailers possessed in, for example, Britain;
(ix)   extension of value-added tax to the retail trade in
        1968 which had the effect of encouraging firms with low
        profit margins but a high turnover;

(x)   the innovative influence of a few individuals.

The initiator of modern trading methods in France was
Edouard Leclerc, the 'grocer of Lardernau' (in Brittany) who,
in 1949, began to undercut his rivals by offering discounts on
the goods he sold.   He attracted considerable opposition from
established shopkeepers anxious to maintain their high profit
margins, but persisted in opening stores in other parts of the
country and won support from the government for his almost-
messianic campaign to bring cheap goods to the people.   The
earliest cut-price stores were simple, unpretentious, with
overheads reduced to the minimum.   But with self-service
methods (introduced to France in 1948), they laid the foundations of the retail revolution that began in the 1960s.
Leclerc, himself, was to be a part of these further changes
and the Centres Leclerc now include 49 hypermarkets and 266
supermarkets.

Leclerc was a forerunner, but the company first responsible for combining his cut-price approach with other American-
type sales methods was Carrefour.   This organization had its
origin in the union in 1959 of two traders in the département
of Ain in eastern France, a small food chain (Badin-Defforey)
and a drapery and fancy goods firm (Fournier).   Together
they opened, in 1960, a small supermarket of 650 sq.m in the
town of Annecy and it was the success of this venture that led
to the decision to build, on American lines, a much larger,
hypermarket, of 4,430 sq.m at Sainte-Geneviève-des-Bois in the
Paris département of Essonne.   In 1982 this pioneering company
was still the largest of its kind in France, responsible for
49 hypermarkets with a sales area of nearly half-a-million
square metres (447,638 sq.m).   Carrefour is followed in size
by Paris-Dock/Mammouth, Euromarché, and Auchan (Metton, 1982).

A hypermarket is defined as a retail outlet having at
least 2,500 sq.m of sales area on a single floor (Dawson, 1976).
A wide range of both food and non-food goods are sold and
these are frequently supplied direct by the manufacturer,
sometimes in special containers that can be employed as display units (Parker, 1975).   Big stocks of goods are held and
the volume of sales is high, assisted by discounts (prix
promotionnels), self-service methods of selection and the use
of a large number of check-out tills.   There is generous car-
parking space which may extend over an area several times
larger than that of the store itself.

Some hypermarkets have a sales area considerably in
excess of 2,500 sq.m.   Of the eleven hypermarkets opened in
the département of Bouches-du-Rhône since 1970, four have a
sales area of between 2,500 and 5,000 sq.m, four of 5/10,000
sq.m and three exceed 10,000 sq.m (Vaudour, 1978).   The
largest is the Carrefour store at Vitrolles with a floorspace
of 21,300 sq.m.   The Carrefour organization operates some of

the biggest hypermarkets in France including the giant store of 25,000 sq.m on the Montauban road north of Toulouse which has parking space for some 4,000 vehicles.

The hypermarket boom reached its peak in the early 1970s and many of the existing stores date from this period. The five hypermarkets in the Toulouse area, for example, were all opened between 1969 and 1972, including the Géant Casino which serves the ZUP of Le Mirail and which opened in 1970 (Idrac, 1979). Three hypermarkets were opened in Orléans within the short space of six months in 1970 offering between them 40,000 sq.m of sales space (Cassou-Mounat, 1978). After 1973 the pace of new construction slowed considerably, partly due to increasing saturation of the market, but partly also to the backlash campaign from small traders discussed below. Sixty-one were opened in 1972 but only 15 in 1975.

Most hypermarkets offer cafeteria facilities and also incorporate a number of small specialist boutiques within the main store as an added attraction. There is often a play area for children and a petrol service station with various auto-related sales. The larger stores may well include separate units devoted to the sale of furniture and garden equipment. In addition, a hypermarket commonly attracts to its general vicinity a range of other shops and services and the proliferation of such outlets, benefiting from the concentration of car-borne shoppers on the locality, has become more common in recent years. Such growth may be encouraged by a suburban commune which derives considerable profit from the taxes paid by the various companies involved (Idrac refers to the million francs paid to the Commune of Portet by Carrefour at Toulouse). By 1979 this Carrefour store had attracted to its vicinity an arcade of 37 small specialist shops, three cinemas, car and caravan sales, furniture stores, a sports shop and garden centre. The trading area of the whole complex, including Carrefour, had grown to 85,000 sq.m (Idrac, 1979). Smaller hypermarkets similarly act as catalysts to the growth of more complex shopping centres, and Soumagne (1977) quotes the example of La Rochelle (population of the agglomeration, 103,000) where hypermarkets were opened in 1970 and 1972. The earlier of these, with 6,000 sq.m of selling area, has attracted some 20 other stores offering furniture, do-it-yourself and other goods so that the area of this enlarged Beaulieu Centre now exceeds 28,000 sq.m.

The ultimate expression of this more comprehensive trading area is the American-type regional shopping centre which incorporates banks, post offices and other high order services more usually associated with city cores. In France such centres are largely confined to Paris where there were ten in 1981, all with a floorspace in excess of 50,000 sq.m. Two of them form the commercial cores of new towns.

French shops traditionally close for a long lunch break

but remain open longer in the evening than their British counterparts. Hypermarkets rely heavily on early evening and weekend trade and Parker (1975) quotes the example of a hypermarket of 18,500 sq.m near Lyon in which 40% of its business is carried out between 6 and 8 p.m. Trade usually peaks on Saturday, however, when the number of customers may well be double that of an average weekday and queues build up around noon even at those stores with a large number of check-outs. The Vitrolles (Marseille) Carrefour averages 18,000 customers on a Saturday compared with an average of 8,000 during the week. Sunday morning trading is most characteristic of the smaller supermarket built to serve the newer suburbs.

The extent of a hypermarket's sphere of influence depends on the store's geographical location as well as its size. A medium-sized hypermarket of 5,000-10,000 sq.m situated in a large city is likely to draw most of its custom from within the city itself, but the big stores (over 20,000 sq.m) can attract customers from as far afield as 50 km. In smaller towns with an established regional function a hypermarket is likely to exert a strong pull over at least 10 km.

The advent of the hypermarket and the general expansion of the commercial sector in the 1960s led to the creation of many new jobs. Beaujeu-Garnier and Bouveret-Gauer give the number for the years 1962 to 1974 as 489,000, equivalent to 19% of total new employment created over that period. But accompanying this overall growth was considerable change within the retailing industry, the arrival of the grandes surfaces having serious consequences for the small independent trader. There were shop closures in the 1950s, but the failure of small businesses gathered momentum through the 1960s with competition, first from the magasins populaires, and later from hypermarkets and supermarkets. Over 100,000 petits commerçants closed between 1962 and 1968 alone (Dyer, 1978), the shrinkage being greatest amongst food retailers and the keepers of small general stores. According to Beaujeu-Garnier and Bouveret-Gauer the number of retailers of specialized food products shrank by no less than 43% between 1961 and 1973. The number of butchers fell by 18% over the same period.

Closure of small shops cannot be attributed entirely to competition from the new stores, though the latter was undoubtedly a major factor. Movement of population away from city centres, coupled with changing domestic habits, has also contributed, as has the generally poor level of organization amongst the independent traders. The geographical distribution of shops tends to be haphazard, especially in the older suburbs, reflecting accidents of history and landownership. Older shopkeepers also lack the necessary kind of professional background to cope with modern forms of distribution and purchasing. In a study of the effect of two hypermarkets (opened in 1973 and 1974) on the established traders in the town of Tarbes,

Palu (1975) placed much emphasis on this characteristic of poor organization. Over a third of the businesses were family-run but, contrary perhaps to general belief, there had been fairly frequent changes of ownership which Palu saw as reflecting an element of instability in the retail situation. Evidence of the variety of trading practices present was to be seen in the 69 variations of opening and closing hours discovered amongst these small shopkeepers.

It was government policy in the late 1960s to encourage the spread of hypermarkets in order to fill gaps in the provision of retail services on the hastily-built estates of grands ensembles. Such support was clearly set out in the guidelines of the Fifth National Plan (1966-1970) and this was sustained during the preparation of the succeeding Sixth Plan (1971-1975) although the latter included some recognition of the problems facing the small independent trader. By 1971-72, however, peak years for the opening of new grandes surfaces, the contraction of the small-scale sector was giving rise to concern and the small shopkeepers had, themselves, found a new champion in Gérard Nicoud, a cafe-owner from the Grenoble area, who believed in 1968-type tactics of direct confrontation. The fruits of his campaign lay in pre-election legislation in 1972 which provided for various kinds of financial aid to shopkeepers including assistance to the elderly wishing to retire. The scheme was funded by a tax on the large stores based on sales area and turnover.

But the real backlash came the following year with the Loi Royer, legislation introduced by the ultra-conservative Minister of Commerce, M.Jean Royer. Behind this Loi d'Orientation du Commerce et de l'Artisanat was a belief that the spread of peripheral hypermarkets ('usines à vendre') had proceeded far enough and that the survival of the small trader was bound up with notions of quality of life and shopping à l'échelle humaine. The legislation thus introduced various measures aimed at supporting the small trader and enabling him to modernize and adapt, imposing at the same time restrictions on the building of new superstores. Control was to be exercised by commissions d'urbanisme commercial, appointed at the level of the département and under the non-voting chairmanship of the departmental préfet. Such committees had, in fact, been set up in embryo in 1969 during the wave of democratization which followed the demonstrations of 1968, but their powers had been little more than consultative and they were strongly influenced by the préfet. Now they had much greater control over any new building that exceeded 1,500 sq.m of selling space (1,000 sq.m in towns with a population of under 40,000) and over extensions of the same scale.

Local trading interests are strongly represented on the 20-member committees which have exercised their restrictive powers with some vigour since 1974. Figures from the Ministry of Commerce, quoted by Beaujeu-Garnier and Bouveret-Gauer, indicate that a greater area of floorspace was refused permission between 1974 and 1977 than was authorized. The committees' powers extend to all types of grande surface, but the refusals were greatest in the case of hypermarkets, more tolerance being shown towards the creation of smaller supermarkets. There is an appeal procedure which enables companies to ask the Ministry of Commerce to over-ride refusal at the departmental level. A number of permissions have been gained in this way, but not sufficient to undermine the role of the committees, which has been seen as an expression of the small shopkeeper's revenge.

Beyond the Hypermarket

As a result of the Loi Royer and the work of its watchdog committees, there have been far fewer hypermarkets opened in France since the mid-1970s than was the case in the early years of that decade. Instead the emphasis has been on the supermarket (which, according to definition, has at least 400 sq.m of selling area) and on the still smaller 'superette' (120-400 sq.m). Many have been built to serve the most recently urbanized areas, especially the new estates of individual houses, and it is these which are well patronized on Sunday mornings. On the estates there is much talk of the virtues of the neighbourhood store as contributing towards the revival of an esprit de quartier.

In city centres the old alimentation générale has given way to specialist outlets, boutiques and to arcades of eye-catching small shops selling luxury, craft and tourist goods. Such changes have usually been associated with plans for restoration of the 'old town', the creation of pedestrian streets and other forms of environmental improvement (Chapter 7). Complementing these schemes of urban renewal have been the redevelopment projects involving the clearance of sites in or close to the urban core and the building of new shop and office complexes. Reference was made in Chapter 3 to major developments like that of La Part-Dieu in Lyon, and the example has been followed on a more modest scale in many other towns. Le Polygone in Montpellier provides a good illustration.

The sociétés de grands magasins have contributed to the success of projects like that of the Polygone by opening department stores which act as a focus in the new centres. By contrast the sociétés have shown little interest in the urban fringe developments. Le Bon Marché opened a store in Sarcelles which failed and the assumption seems to have been that the future of the department store continues to lie with the wealthier clientele shopping in the city centre. An exception is to be found in the case of the multi-functional centres built in the capital's better-off suburbs. The first of these, Parly Deux near Versailles, opened in 1969 and offers

58,000 sq.m of sales area.   Vélizy Deux (1972) is even larger
with 84,000 sq.m.   Both are in the wealthier, south-western
suburbs, in contrast with Sarcelles to the north of the city,
where the corresponding centre of Les Flanades has experienced
great difficulties, and care has been taken to create attractive 'shopping environments' with spacious malls, fountains
and the use of a wide range of construction materials (Smith,
1973).   Each of the centres has a couple of large department
stores which, in the modern manner, terminate the malls.

It has been suggested that the specialist centre will
emerge over the next decade as a significant type in France
(Dawson, 1981).   Specialist stores, particularly those selling
furniture, have been found on the edges of French cities for
many years and Smith sees these as having established a prece-
dent for the development of hypermarkets.   To the furniture
stores have been added more recently many other kinds of
specialist shops and the main radial roads leading out of most
towns now present a succession of car showrooms, caravan sales,
stores selling electrical goods, kitchen ware, camping and
sports equipment, garden centres, show houses and even night
clubs and discos.   All cater for the car-borne customer and
roadside advertising is prominent.   Such proliferation of
surfaces de vente may herald the emergence of the specialist
centre.   At present its form tends to be linear, straggling
and accompanied by much that is brash and vulgar, contributing
to what has been described as 'la chienlit urbanistique'.
But it is likely that as the novelty wears off attempts will
be made to tidy up the worst of the sprawl and to group the
traders on sales parks.

Chapter 7.

RESTORATION OF THE CITY CENTRE

Hypermarkets and other surfaces de vente are just one
element in the 'exurbanisation' that has become such a feature
of French cities over about the last 15 years. Manufacturing
industry, for example, has shown a strong tendency to move
out to the suburbs in search of cheaper sites and access to
modern highways. It is a trend that has been encouraged by
the creation there of zones industrielles, special financial
provisions for which were made under the Fifth Plan (1966-70).
More recent and more elaborate than the ZIs are the parcs
industriels where manufacturing enterprises are joined by a
variety of other activities including distribution, banking
and other services for the industrial community (Barrère
and Cassou-Mounat, 1980). Wholesale food markets and
exhibition centres compete with these industrial estates for
sites in the urban fringe.

Lycées and colleges have also abandoned outdated buildings
in the city centre for more spacious surroundings further out.
The phenomenon is even more marked in the case of universities,
however, vast campuses having replaced older faculty buildings
that had become increasingly crowded under the pressure of
growing student numbers. The university of Caen was a forerunner, but this example was quickly followed by others whose
campuses have attracted staff housing, services and sports
facilities, leading to the creation of what are virtually
separate university suburbs. The campuses are attractively
landscaped but some regret the loss of intimacy associated with
the narrow streets, cafe life and other amenities of the city
core.

Colonization of the urban fringe by superstores, factories
and colleges has increased the value of surrounding building
land but the effect is uneven, depending on the attractiveness
or otherwise of these neighbouring sites for residential use.
This, coupled with the way in which the housing market operates,
has encouraged social segregation. Jeanneau (1974), for
example, in a study of the urban fringe of Angers, observes

that, 'le contraste entre quartiers de petits salariés et
quartiers de résidence aisée est plus marqué sur la périphérie
que dans le centre'. The contrasts are most obvious between
the ZUPs with their concentrations of HLM housing, the villas
of the salaried classes on higher ground overlooking the
campuses, and the estates of box-like pavillons in their little
garden plots.

The new developments, both residential and non-residential,
tend to be prodigal in their use of space and have extended
greatly the built-up area of French cities. Jeanneau (1978)
quotes the example of Cholet, the urbanized area of which grew
from 400 ha. in 1962 to 800 ha. in 1968, doubling again to
1,600 ha. in 1977. Physical sprawl is most evident in those
towns, like Cholet, which have experienced rapid population
growth (2-3% a year since 1962 in this case) but the phenomenon
is universal as a result of the shift of people and activities
from centre to periphery. Surrounding the cities is also a
zone of 'rurbanisation' in which villages have become dormitory settlements and, further out, favoured places for weekend retreats. Geoffroy (1980), using the examples of Laval,
Niort and Le Roche-sur-Yon, suggests that this process of
rurbanisation is evident to a distance of some 10 km from the
centre of a medium-sized town, and the zone is clearly much
wider in the case of larger cities. It has the effect of
extending the physical impact of the growing city beyond, not
only the municipal boundary, but also the limits of the area
for which a plan (schéma directeur) is being prepared. The
consequences for land use planning are obvious, and in a study
of the effect of the growth of Orléans on its surrounding
region, Vassal (1977) draws attention to 'la rupture de
l'équilibre ville-campagne'.

Urbanization of the fringe is unquestionably the most
striking aspect of the postwar transformation of French cities.
Old town cores are now embedded in a sprawling matrix of
residential, commercial and industrial land uses loosely knit
by a web of new highways - pénétrants, rocades and boulevards
périphériques - which together have totally altered the scale
at which one views the city. 'Les périphéries urbaines
deviennent un chantier de créations commerciales effrénées'
(Metton, 1982). But the urban cores, themselves, have also
experienced change, partly as a result of the movement out of
the centre of many activities, partly in response to planned
redevelopment and urban renewal. Latterly there has been an
increasing emphasis on the restoration and rehabilitation of
old buildings of historical and architectural importance.
Most large cities have undertaken at least one major project
of urban renewal and examples of these were described in the
chapter (3) devoted to the métropoles d'équilibre. The
present chapter is concerned with the complementary theme of
renovation.

Impoverishment of the Core

Accompanying the transfer of activities from core to periphery has been a loss of population from the central quartiers of most French cities. Poitiers may serve as an example (Figure 15). Here the well-defined central area lost more than a quarter of its population between 1962 and 1975, the total falling from 23,260 at the former date to little more than 17,000 by the mid-1970s. The rate of loss also accelerated over the period in question, giving rise to concern for the future of once close-knit local communities and their traditional vie du quartier. Not only were their numbers shrinking, but those who remained in the centre were finding it increasingly necessary to travel to the suburbs for medical, educational and other social services which had once been provided locally and within walking distance of their homes (Pitié, 1979).

The experience of Poitiers has been repeated throughout France. Removal of families to the suburbs has been partly offset by the arrival of immigrants from North Africa or Portugal and by the usual transient population of the young and the bohemian. But the overall trend has been towards an increasingly elderly, poor and, in the early 1980s, unemployed population. The number of vacant dwellings has grown and historic buildings have been allowed to decay for want of funds or municipal resolve. Words like paupérisation and dégénérescence have been used to describe the demographic and architectural state of city centres. Writing of the situation before the first serious attempts were made at restoration, Paul Boury, deputy mayor of Dijon, put it plainly, if somewhat dramatically, when he said, 'le centre de la ville était abandonné, il allait mourir; certains parlaient déjà du "crépuscule des villes", de " l'agonie des centres" ' (Le Monde, 5 March, 1980).

In France, as in Britain, the response to central city decay was initially one of clearance and redevelopment. In some cases, as in Lyon (La Part-Dieu) and Montpellier (Le Polygone), the sites for redevelopment were provided by former military barracks, but most schemes involved the demolition of selected îlots of mixed land use declared insanitary and beyond repair by virtue of delapidation and lack of basic amenities. The operations were traumatic for residents who were moved to tower blocks on the edge of town. Barrère and Cassou-Mounat (1980) note that the initial intention may have been to rehouse the local population, but plans were frequently changed and rehousing projects gave way to centres directionnels of administrative and private offices, shopping arcades and luxury apartments. Slowness to redevelop was due to problems of acquiring land in the face of local opposition, to the cost of compensation and, sometimes, to doubts over which was the right solution (Les Halles).

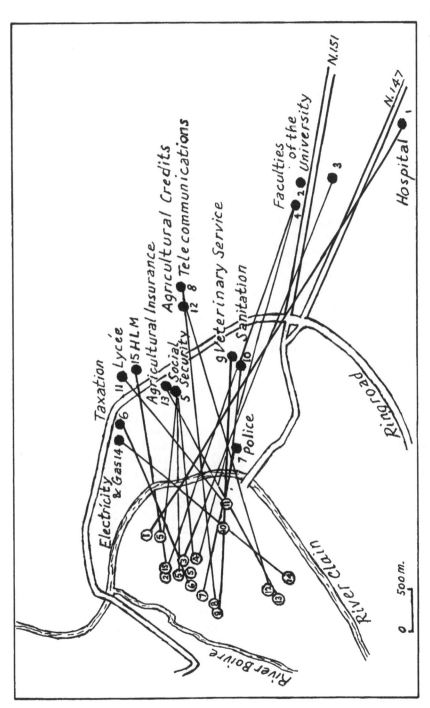

Figure 15. Movement of public services from the centre of Poitiers to the suburbs, after Pitié, 1979

The vogue for clearance was at its height in the late 1950s and early 1960s. Many old buildings were declared to be slums and demolished in favour of new ones whose bulk and the materials used in their construction rendered them wholly unsympathetic to their surroundings. The treatment of Paris was described in Anthony Sutcliffe's The Autumn of Central Paris, but smaller towns did not escape the bulldozer. One which suffered badly from this kind of treatment was Metz where, between 1958 and 1965, numerous operations were carried out. These involved the loss in one quartier of houses as old as the thirteenth century, cleared to make way for a bus station. The changes were first catalogued by Jean de Mousson in an article in Le Lorrain (1 November, 1966) headed, 'Le vieux Metz méritait un sort meilleur qu'une destruction systématique', and a lengthy description followed in Le Monde (30 April, 1970). The author of the latter, Philippe Levantal, concluded: 'De l'histoire de Metz, restera-t-il autre chose qu'un musée de fragments? Voilà donc ce que l'on fait d'une ville d'art. Ces démolitions, practiquées au coup par coup, au gré des expropriations, sans plan d'urbanisme cohérent, relèvent en fait d'un "haussmannisme" attardé'.

Many other towns experienced misfortunes similar to those of Metz and it was the publicity given to this kind of clean-sweep planning which gradually awakened public opinion to its consequences and brought about a change of policy in favour of restoration and rehabilitation.

Restoration

The practice of listing monuments historiques was established in France as early as 1837 and certain planning powers aimed at protecting listed buildings were introduced in 1913. The weakness of the legislation, however, was that it dealt only with individual buildings and not with their surroundings so that a seventeenth-century hôtel or a romanesque church might be left standing in a wasteland of demolition, later to be swallowed up in some development of totally different architectural character. An Act of May, 1930 had provided for what was called the preservation of sites but this was of benefit only to villages and such small towns as Conques, Les Baux and Pézenas which had an accepted reputation and attracted many visitors. Hence the importance of the Act introduced by M.André Malraux, Minister of Cultural Affairs, in August, 1962.

The intention of the 'Loi Malraux' was to safeguard historic urban cores from the kind of insensitive redevelopment that was widespread at the time. It did this by introducing the secteur sauvegardé, the aim of which was to protect whole areas rather than individual monuments. It recognized that the historic quarters of towns and cities derive their character from their environmental setting - the street scene,

trees and open spaces - as well as from the buildings that
separately comprise it.   In this sense the Act was a forerunner of the conservation area, introduced in Britain
following the Civic Amenities Act of 1967, and it also influenced conservation policies in other countries.   The Loi Malraux
sought to encourage restoration as well as giving protection
but, as we shall see, the procedures introduced have proved
ponderous and slow, and the principal merit of the legislation
must be seen in the change in attitude to the historic town
centre which it brought about.

A secteur sauvegardé is designated after consultation
between the ministries concerned and the local authorities and
a plan for its preservation and restoration is then drawn up
by an architect appointed by the mayor on the recommendation
of the Commission Nationale des Secteurs Sauvegardés.   Funds
then become available from the Crédit Foncier, mainly in the
form of loans, which can amount to two-thirds of the cost of
the operations envisaged.   It has been usual for the actual
work of restoration to be entrusted to a specially-appointed
mixed economy company which selects one or more operational
areas (îlots opérationnels) on which to concentrate its
activities.   Secteurs sauvegardés vary widely in size from
well over 100 ha. (e.g. Bordeaux 150, Nancy 140) to under ten
hectares (Avignon 6) and within these the size of the operational area averaged a little less than two hectares in the
early years of the scheme.   By the 1970s it had become
smaller, closer to one hectare as costs had risen.

There was much initial interest in the possibilities of
the Loi Malraux and the first ten secteurs sauvegardés were
designated in 1964.   There were 35 by the end of 1968;
thereafter fewer were delimited and the total had reached only
60 by 1976 (Figure 16).   No additions were made to the list
until 1981 when Viviers (Ardèche) became the first town for
five years to declare a secteur sauvegardé.   The earliest of
all was that of Le Vieux Lyon where 30 ha. of the old town
were designated a secteur sauvegardé in May, 1964.   This may
serve as an example of the way in which the provisions of the
Loi Malraux have been utilized.

Old Lyon lies between the slopes of Fourvière and the
Saône and it owes its architectural character to the wealth
of ornate, renaissance buildings, many dating from the sixteenth century when Lyon became an important banking centre
although their foundations often predate this period.
Individual houses are tall and narrow, frequently reaching
five storeys, and they have interior courtyards linked by
covered passageways, known as traboules.   Stone doorways with
heraldic emblems above them look out on narrow, paved streets.
Physical decay began when its wealthy bourgeois residents
later abandoned the area for newly-fashionable quarters built
on the peninsula between the Saône and the Rhône, and by the

Figure 16. Towns with a secteur sauvegardé, 1980

late nineteenth century these old quarters housed a mainly
poor population of labourers and small traders who subdivided
the houses, converting some into workshops.   Fortunately for
future restoration the poverty of the inhabitants led to neglect,
and although individual buildings have suffered greatly from
lack of care and from alterations, the area was still relatively intact in the early 1960s.

Concern was being expressed for Old Lyon before 1962,
notably by an amenity society, 'Renaissance du Vieux Lyon',
founded in 1948, and by bodies such as the Junior Chamber of
Commerce which helped to organize fairs and other events in
order to draw attention to the attractions and the needs of
the district.   In 1963 the first national conference on the
subject of 'Les Quartiers Anciens' was held in Lyon and this
was followed by the setting up of a mixed economy company, La
Société d'Economie Mixte pour la Restoration du Vieux-Lyon
(SEMIRELY), to take advantage of the Malraux Act.

The secteur sauvegardé that was delimited covers approximately 30 ha. in a narrow strip running north to south and
housed a mixed population of some 15,000 including Spanish,
Portuguese and other immigrants.   Within this, two operational blocks were agreed upon, together accounting for about one
hectare of land in the heart of the secteur and housing 1,600
people of whom 82% were classed as 'poor', 14% as small shopkeepers, and only 4% as of the middle or upper classes.
Seventy-three per cent were French, the rest immigrant.   Some
idea of the social problems accompanying those of restoration
may be gauged from the fact that the 47 buildings involved in
the operation were divided in 1964 into no fewer than 419
apartments, 149 furnished rooms, 3 dormitories for immigrant
workers and 50 shops.   Most living units were rented but
many of these were in buildings that were under joint-owner-
ship.   It was anticipated that, following restoration, no
more than 269 modernized apartments would remain and that some
families would therefore have to move elsewhere.

It is scarcely surprising in view of the above that
restoration work should have been slow.   Architecturally it
has proved very successful, and the rate of improvement
increased as the area once more became an attractive place of
residence, but it has been accompanied by considerable social
change.   Such change began with the rehousing of many of the
tenant families in HLM flats in other parts of the city and
the sale of properties to the mixed economy company by owners
who were too old or poor to face the prospect of repaying
loans after restoration had been completed.   Some buildings
were, in fact, bought as an investment by outsiders who gained
from the loans available and had time to wait for the work of
improvement to be completed.   A special loan fund was also
set up by the City Council to help pay for the restoration of
shop fronts not covered by the Malraux legislation.   Following

restoration, the rents charged by owners tend to be many times higher than those paid previously. In Lyon - and the same was true of other towns in the early years - there was no policy of letting properties within the area to former residents and no subsidies were available to enable former tenants to return. In fact under a fifth of the original population has stayed. The result has been gentrification, and there is little reason to think that this was viewed with dismay by the local authority.

What has happened in Lyon is not dissimilar from that which has been experienced elsewhere. One of the best documented, and perhaps the most controversial, examples of restoration following the Malraux Act has been that of Le Marais in Paris where a secteur sauvegardé of 126 ha. was designated in April, 1965. Seventeenth-century palais and hôtels have been painstakingly restored but often at the expense of craft workshops which had colonized their gardens and courtyards in the nineteenth century. Old shops have also given way to bistrots, boutiques and stores selling antiques and other expensive articles. Critics are inclined to view the operation as the creation of an architectural museum-piece and express regret at the loss of animation and of the district's identity and sense of place. Chaline (1980) has used the term pigalisation for the kind of up-market transformation that has taken place in Le Marais and the safeguarded quarters of other cities.

Designation of a secteur sauvegardé in a small town has released a source of funds for a programme of conservation that would otherwise have been beyond the capacity of the local authority concerned. Examples are to be found amongst the castle towns of the Loire valley where secteurs sauvegardés were declared in Chinon and Loches in 1968. Both towns have a resident population of less than 10,000 and without assistance could not have undertaken the costly work of restoration that has been carried out around the Grand Carroi and in the Cité Médiévale. Nearby Richelieu, with fewer than 3,000 inhabitants, has similarly been able to restore houses along the Grande Rue built when the Cardinal laid out his planned town in the early seventeenth century.

A somewhat larger town, but one that has been obliged to undertake a formidable task of restoration, is Saumur where the population in 1975 totalled 32,500. The municipal council of Saumur was one of the first to take advantage of the Loi Malraux and a secteur sauvegardé of 32 ha. was designated in August, 1964 (Figure 17). Within this, an operational area of 2.6 ha. was defined which took in the oldest buildings, some of them of the fifteenth century, clustering at the foot of the castle hill around the Place Saint-Pierre and along the Grand'Rue. The work of restoration was entrusted to a mixed economy company which, by 1976, had completed its task on 42

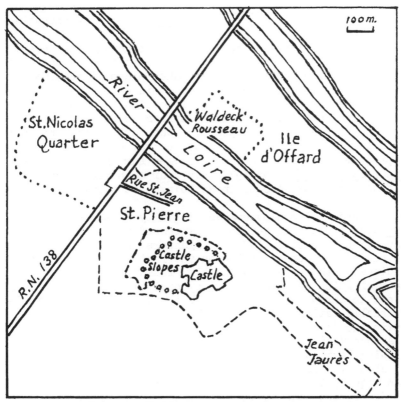

100m.

River

St.Nicolas
Quarter

Waldeck
Rousseau

Ile
d'Offard

Loire

Rue St. Jean

St. Pierre

R.N. 138

Castle
Slopes

Castle

Jean
Jaurès

_ _ _ _ Extent of the secteur sauvegardé
_._._. Limit of the operational area
.... Quarters undergoing rehabilitation
o o o o Urban renewal on castle slopes

Figure 17. Restoration and rehabilitation in Saumur, after
            Jeanneau, 1978

buildings, with repairs in progress on 13 others (Jeanneau,
1978).   Additional funds have been available since 1976 when
Saumur obtained a <u>contrat ville moyenne</u> and these have been
devoted to complementary works involving pedestrianized streets
in other parts of the <u>secteur sauvegardé</u> and the improvement
of dwelling houses in adjacent <u>quartiers</u> where the emphasis is
on rehabilitation.

Having experience of restoration extending over some 20
years, Saumur also affords a useful illustration of the change
in approach to this work that has taken place since M.Malraux
introduced his legislation.   Here, as in other towns, improvement has been accompanied by gentrification and the programme
has attracted criticism for its slowness, cost, and its emphasis
on the most prestigious projects.   Such criticisms are reflected in the length of time taken by central government to approve
the programme of conservation envisaged for the whole of a
<u>secteur sauvegardé</u>.   By 1977 only four such programmes had
won final approval, one of these being that submitted for
Saumur.

Concern at the slowness and regidity of the <u>secteur
sauvegardé</u> procedures was being expressed with increasing
frequency by the early 1970s.   Planners and architects were
also blamed for concentrating on the restoration of historic
buildings at the expense of humbler residential property, the
repair of which was urgent if the population of city centres
was not to shrink further.   The Loi Malraux had made provision
for rehabilitation work of this kind but progress was slow
despite the setting up of a special body to fund improvement,
the Agence Nationale pour l'Amélioration de l'Habitat (ANAH),
following legislation in 1970.   In response to the disquiet
felt about the condition of older housing the government set
up an enquiry in 1975 under M.Simon Nora and in its report the
Nora Commission made strong recommendations in favour of the
modernization of such housing, linking this with the need to
improve amenities and restore life to city centres.   Improvement at the centre was seen as an alternative to building soul-
less estates on the periphery and local authorities were urged
to take greater initiative in the rehabilitation of older property.   The Nora Commission's suggestions bore immediate fruit
with the establishment in 1976 of the Fonds d'Aménagement Urbain
(FAU) and the introduction of <u>opérations programmées
d'amélioration de l'habitat</u> (OPAH).   The need to improve the
existing stock of dwellings was also stressed in the Seventh
National Plan (1976-80) which included a <u>programme d'action
prioritaire</u> under the title of 'Mieux Vivre dans la Ville'.

The FAU was set up to coordinate the activities of ministries and departments already engaged in the work of restoration
and urban renewal.   It has offices in the <u>départements</u> as well
as Paris and it has a wide brief to use its funds for purposes
as diverse as the pedestrianization of streets, the creation of

open spaces, the restoration of historic buildings and the upgrading of HLM apartments. The creation of FAU was intended to introduce greater flexibility to the urban improvement programme and to reduce the degree of central control over this work. It had immediate implications for the politique des villes moyennes (Chapter 4) as well as for the way in which restoration work was financed in the secteurs sauvegardés.

The FAU was established with the aim, not only of financing repairs to property, but also of improving the 'quality of life' and living conditions of property owners with limited means. This social objective is evident also in the OPAH procedure, operational from 1977. The OPAH differs from the secteur sauvegardé in several important respects. There is much greater emphasis, for example, on local agreement and cooperation and every effort is made to retain the already-resident population in the area. Localities chosen for an OPAH are not restricted to those possessing the most historic buildings, the aim being to rehabilitate more modest property in districts that have become run-down in order to benefit local owners and tenants. In this respect there are parallels with the General Improvement Area legislation introduced in the United Kingdom in 1969. The actual work of improvement is directed by a locally-appointed body, often a mixed economy company, and funds are drawn from the ANAH and the FAU. Significantly these include rent subsidies offered to families who would not otherwise be able to remain in the area.

The OPAH is successor to the îlot opérationnel, the last of which was created in the Petite Venise quarter of Colmar in 1975. Some 500 OPAH projects were begun in the first five years of the scheme. They range in size from a few streets to the entire centre of a small town and the strategy is also used for village improvement in rural areas. There is a 3-year limit to the programme proposed in order to avoid the long delays that have been common in the secteurs sauvegardés. Some towns which already had a secteur sauvegardé have also used the OPAH procedure in order to extend the work of restoration and improvement to other parts of the central area or the twilight zones of the 'inner city'. Such is the case at Saumur.

The Saint-Nicolas quarter of Saumur, on the fringe of the historic core of the town, had lost a third of its resident population between 1962 and 1975 and was zoned for commercial redevelopment. As an alternative to this solution, however, it was decided to declare an OPAH in Saint-Nicolas and rehabilitate as many as possible of the existing 574 dwellings in the quarter, providing better social amenities there as well as some of the earlier-proposed commercial development (Kain, 1982). Similar improvement work has subsequently been extended to other quarters on the edge of the urban core: Waldeck-Rousseau and Jean-Jaurès (Figure 17). In addition to these,

another project involving urban renewal has been carried out
on the slopes of the hill below the château and in the heart
of the secteur sauvegardé.   This has included slope stabilization and the clearance of ruinous structures of little
architectural merit.   In their place a new development has
been carried out since 1977 involving 280 new dwellings
grouped in what has been described by the architects concerned
as being in 'the traditional Loire valley village style'.
Care has been given to landscaping with, for example, all
electricity and telephone wires put underground.   A hundred
of the units are HLM dwellings, indicative of the social
objectives that are now stressed in these improvement schemes,
and the project has had the effect of re-introducing about 500
people to the Quartier du Coteau.

Twenty years of conservation have brought life and beauty
back to the centres of French cities.   In 1962 the Loi
Malraux alerted public opinion to the consequences of further
neglect and saved from demolition whole districts of historic
towns, like the quartier des Tanneurs in Colmar with its half-
timbered houses and narrow paved lanes which had been largely
emptied of its population in preparation for the bulldozer but
which was saved and subsequently restored.   The emphasis in
the 1960s on the architectural and the aesthetic resulted,
however, in the displacement of older and poorer residents
in favour of a trendier population employed in the offices of
the nearby centres directionnels and wealthy enough to shop
in the luxury boutiques which had succeeded the more traditional shops in the newly-restored streets.   But greater
efforts have been made in the last ten years or so to retain
the local population, and there is interest in rebuilding
something of the community life, the esprit de quartier, that
was typical of these central city districts.

The further revival of city centres depends to a large
extent on finding a solution to problems of traffic circulation.   Schemes for controlling the movement of the private
car and for assisting public transport have been slow to be
adopted and the need for traffic management has been widely
acknowledged only in the 1970s.   A number of pedestrian
streets were incorporated in the plans for towns rebuilt
following war damage, but the first serious attempt to introduce pedestrianization to an existing city centre was the
operation completed in Rouen's rue du Gros-Horloge as recently
as 1972 (Metton and Meynier, 1981).   Since that time, however,
the vogue for pedestrian streets has grown rapidly as traders
have lost their earlier hesitation and recognized the advantages that these can bring.   Adopted first in the larger
cities, pedestrianization spread to the medium-sized towns
where plans for one or more rues piétonnières were often
included in contrats villes moyennes.   Now such towns as
Rodez, La Rochelle and Manosque can boast a system of pedes-

trian ways to complement the work of restoration that has been carried out in their historic cores. In such places 'la plateau piétonnier remplace la rue commerçante' (Palu, 1982). But not all have been so fortunate, or so progressive, and some towns, of which Dinan in Brittany may serve as an example, are still beset by severe traffic problems.

The centres of most French cities exhibit sharp contrasts over relatively short distances. Blocks of decaying properties stand close to the cleaned and restored facades of seventeenth-century hôtels; the overcrowded dwellings of immigrant families are to be found within a stone's throw of the gentrified apartments of young cadres (professional); smart new office blocks look down on abandoned factory sites; whilst the antique and souvenir shops of a newly-cobbled rue piéton have little in common with the back street câfé-bar. Such variety of land use defies the search for a simple model of city centre structure. Rapid change is inhibited by the nature of land ownership and land value as well as by the constraints imposed by the existing fabric of streets and buildings. The direction of such change will continue to depend to a large extent on the policies determined by national government and the will of local authorities to apply them.

Chapter 8.

PARIS

It seems likely that the population of the Paris Region (known officially since 1976 as Ile-de-France) reached ten millions some time in the course of 1980. The census of March, 1982 recorded a provisional total of 10,056,100. This represents 18.6% of the population of France on 2.2% of the national territory. In 1954 the corresponding proportion was 17.1%.

Since 1946 the population of the Region has grown by some three-and-a-half millions (Table 9). Until recently the rate of population growth has exceeded that of France as a whole, averaging 1.90% per annum between 1954 and 1962 compared with the national figure of 1.09%, 1.46% between 1962 and 1968 (France 1.14%) and 0.97% between 1968 and 1975 (France 0.81%). Growth has been sustained by a rate of natural increase that has been above the national average and, in the earlier part of the period, by higher than average rates of inward migration (Table 4, Chapter 2). By 1968-75, however, the annual gain attributable to migration (0.17%) was below the national figure (0.23%). Since 1975 the annual rate of growth of the Paris Region's population (0.3%) has actually fallen below the national average (0.4%). Natural increase added an average of 67,000 a year to the Region's population between 1975 and 1982 but there was now a migratory deficit which amounted to -42,000 a year over the period.

This change in demographic fortunes is a consequence, in part, of the nationally-falling birth rate and of the halt placed on migration from overseas. But there is clearly a regional component as well, represented in the position of Paris relative to the rest of France. Between 1954 and 1962, for example, the Paris Region received each year an average of 43,000 more migrants from other parts of France than it sent back to the provinces. By 1962-68, however, the net gain had fallen to little more than 11,000 and in the succeeding intercensal period there was a net loss of around 20,000 a year in this internal migratory movement. The fact

TABLE 9. POPULATION CHANGE AND REDISTRIBUTION IN THE PARIS REGION
(Population total in thousands)

	1946	1954	1962	1968	1975	1982
Ile-de-France	6,597.9	7,317.2	8,470.0	9,248.6	9,878.6	10,056.1
PARIS	41.3%	39.0%	33.0%	28.0%	23.3%	21.6%
Petite Couronne	36.3%	37.3%	40.6%	41.4%	40.3%	38.8%
Grande Couronne	22.4%	23.7%	26.4%	30.6%	36.4%	39.6%

Source: INSEE, July, 1982

that there was still an overall increase of population due to
migration between 1968 and 1975 is due to the greater number
of foreign-born entering Paris than leaving (+35,000 a year).
Following the ending of labour recruitment abroad, this latter
component was no longer sufficient by the late 1970s to offset the negative internal flow.

Clearly Paris no longer dominates the growth of the French
population as it has for over a century and a half.   Many
reasons have been suggested for this change including the high
cost of rent and property, the expense and tedium of long
journeys to work and other diseconomies.   Government sources
prefer to see it as a measure of the success of policies aimed
at decentralization.

Hidden behind the totals for Ile-de-France as a whole, a
very considerable redistribution of the population has been
taking place within the Paris Region over the course of the
last 30 years.   The core of the Region, the département of
Paris which corresponds with the 20 arrondissements of the old
Cité, is now much less dominant in population terms than it
was in the early 1950s.   In 1954 it accounted for two-fifths
(39.0%) of the Region's total, but this had fallen to little
more than one-fifth (21.6%) by 1982 (Table 9).   Over the same
period the proportion of the Region's population living in the
three départements (Hauts-de-Seine, Seine-Saint-Denis, Val-de-
Marne) of the 'Petite Couronne' - usually referred to as the
inner suburbs - has risen very slightly from 36.3% of the whole
to 38.8%, whilst that of the outer suburbs of the 'Grande
Couronne' (départements of Essonne, Seine-et-Marne, Val-d'Oise,
Yvelines) has almost doubled, from 22.4% to 39.6%.

The City of Paris recorded a postwar population peak of
2,850,000 in 1954.   Thereafter the total fell by an average
of 1.2% per annum between 1962 and 1968 and at the faster
rate of 1.6% annually between 1968 and 1975.   By 1975 the
total had fallen to just under 2.3 millions, the City having
lost almost a fifth of its population in twenty years.   The
political implications of this loss are considerable.
M.Chirac, a leading contender for the presidency, became mayor
of Paris in 1977, and the provisional results of the 1982
census were thus awaited with more than usual interest.
Announced in September of that year, they showed that the
drift of population had continued, but at a slower rate, the
annual loss amounting to 0.8%, only half that of the preceding
seven years.   The total population of the City of Paris in
March, 1982 was 2,168,300.

A study of individual arrondissements and quartiers
carried out by the Atelier Parisien d'Urbanisme suggested that
schemes of urban renewal had had the greatest single influence
on population change.   Losses had been heaviest in the central
arrondissements, especially the 1st, 2nd and 4th, and in those
that had witnessed major office developments (the 8th

136

arrondissement, the quartier Chaillot, the 16th and the
Chaussée-d'Antin and Faubourg-Montmartre quartiers in the 9th
arrondissement).   Elsewhere much depended on the extent to
which the existing housing stock had been renewed or
rehabilitated.   Where little had been done, the loss of
population was above average, but in some quartiers there had
been an increase following the completion of urban renewal
programmes (quartier de la gare in the 13th arrondissement,
quartier de la Villette, Pont de Flandres et d'Amérique in
the 19th).   The study also noted that the losses had been
greatest amongst the least-well-off and younger sections of
the population, especially of families with children.   The
trend in the 1970s has been towards 'embourgeoisement' and,
because there are so many households (60%) without children,
'vieillissement' in the City.   Another characteristic is the
high proportion of foreign-born, a fifth of the population
being classed as migrant.
        Most of the population growth recorded in the départements
of the Petite Couronne took place before 1968 when the inner
suburbs housed 41.4% of the total of the Paris Region.   Since
that date the proportion has fallen and these inner suburbs
have been experiencing a net loss of population by migration
throughout the 1970s.   Between 1975 and 1982 their population as a whole fell by 77,000.   The outer départements,
by contrast, have witnessed continuous growth as families have
moved out to the grands ensembles, the new towns and the
estates of pavillon housing built on the urban fringe and
served by motorways or extensions to the métro system.   The
four départements recorded an average net addition of 110,000
persons a year between 1968 and 1975.   For one of them,
Essonne, this meant an overall increase in excess of 37% with
all its attendant demands for new services.   There was
further growth totalling 386,000 between 1975 and 1982.
        Additional evidence of this spreading out of greater
Paris may be found in the fact that the Parisian agglomération
suffered a small reduction in population between 1975 and
1982 from 8,549,000 to 8,505,000, whilst the peripheral villes
moyennes of Ile-de-France, such as Melun and Mantes, experienced high growth rates.   This suggests that a part at least of
the recent negative migratory balance between the Paris Region
and the provinces (above) may be accounted for by the fact
that the capital's dormitory ring now extends beyond even the
boundaries of Ile-de-France.

Morphology and Housing
        Land use in the City of Paris exhibits great diversity
but a clear distinction can be made between the historic core,
which corresponds broadly with the first nine arrondissements,
and the periphery.   Within the former there is clustering of
related activities in distinct districts.   On the left bank,

the administrative quarter occupies the 7th arrondissement but
extends also into the 15th, whilst the university and intellectual life of Paris is closely identified with the 5th and 6th
arrondissements.   Here the schools that were to become the
University of Paris first grew up around the religious houses
which occupied the slopes of Mont-Sainte-Geneviève.   Across
the Seine, the main shopping streets and fashion houses of the
first four arrondissements give way westwards to an office
quarter in the 8th and 9th which displays an intricate and
rather incoherent pattern of land occupancy in a wide range of
business premises.

Surrounding this core are the mixed industrial and residential districts of the peripheral arrondissements (11-20).
Growth took place principally during the second half of the
nineteenth century and in the years up to the First World War
when light industries followed the railways and canals.   Housing
is of poor quality, built close to the factory or workshop,
and there is little open space.   Traffic clogs the narrow
streets and, according to Ardagh (1977) 'even a poorish London
district like Battersea or Lambeth seems a haven of space and
calm' in comparison with the crowded quartiers of, say, the
10th, 11th or 13th arrondissements.   Pinchemel (1979) refers
to an industrial proletariat huddled together under exceptional
conditions of poverty, and quotes a description of the urban
landscape in a report to the Schéma Directeur (1976):
'désarticulé, son paysage désordonné, inachevé'.   These
working class areas are most typical of eastern Paris and are
in total contrast with the wealthy suburbs of the 16th
arrondissement which border the Bois de Boulogne in the west
of the City.

It is in these peripheral arrondissements that some of the
biggest urban renewal schemes have been carried out since 1960.
Houses, workshops and warehouses have been cleared to make way
for major new office complexes and apartment blocks.   Examples
are the well-publicized developments at Fronts-de-Seine, Maine-
Montparnasse, Place d'Italie and Bercy-Gare de Lyon.   Critics,
like Sutcliffe (1970), resent the intrusion of these concrete
towers into the historic skyline, whilst others regret the loss
of sociability associated with the shared poverty of the slum
and the street.

The population of Paris was largely accommodated within
the confines of the walled City until the final quarter of the
nineteenth century.   Even then the growth of the suburbs was
relatively slow and rapid expansion came only after the First
World War.   One of the factors encouraging suburbanization
was the continued existence of the octroi, a tax on goods which
was levied on their entrance to the City.   This raised the
price of basic commodities, including food, and to escape it
people moved to communes outside the City limits - Vanves or
Malakoff in the south, Aubervilliers or Pantin in the north -

where they lived under conditions that were often as crowded as those within Paris itself. Some settled in the zone that had been reserved for military use when the walls were built in 1845 and these were later rehoused in cheap apartments known as habitations à bon marché (HBM).

Suburbanization was also brought about by the expansion of manufacturing industry and the setting up of new factories for cars, aircraft, electrical equipment and chemicals. Some of the largest were on the gravel terraces that bordered the Seine downstream from the centre. The eastern suburbs attracted fewer large employers and were largely given over to poorly-planned and ill-equipped housing for a working-class population many of whom were obliged to make long journeys to their place of employment. Hall (1966) has described these suburbs as 'a vast, ill-conceived, hastily-constructed emergency camp to house the labour of Paris'. It is 'la triste banlieu', with many deprived families, and support for the Parti Communiste has always been strong, hence the other popular title of 'the red belt' for these inner suburbs of the Petite Couronne. The special needs of this zone have been recognized in successive plans for greater Paris and have resulted in some major schemes of urban renewal, at La Défense for example, and also at Créteil and Bobigny, both now the administrative centres of their respective départements (below).

Growth of the suburbs following the First World War was assisted by the introduction of the 8-hour day and by extensions made to the rail network. It was also brought about by speculative housing development involving the sub-division of farm holdings by lotisseurs. Between 1920 and 1930 alone, some 9,000 hectares were subdivided into 250,000 parcels for building purposes according to Pinchemel. Tiny, geometrically-shaped plots were created for dwellings that were often little better than shanties. There was some tendency for them to follow the mainline railways but otherwise their distribution was unplanned. Sites chosen were frequently unsuitable, reflecting only the underlying pattern of land ownership and the skill of the speculator in acquiring plots. Most suffered from poor access to services, including schools and hospitals, and special legislation in the form of the 'Loi Sarraut' of March, 1928 was required in order to ensure that the mal-lotis were provided with such basic needs as water, drainage, roads and electricity (Bastié, 1964 a).

In her study of change in Paris between 1878 and 1978, Norma Evenson (1979) attributes the popularity of this kind of development to the Parisian artisan's preference, given the chance, for a single house on its individual plot where he can express his individuality and his own taste, however eccentric. Opportunities for self-expression were certainly plentiful where dwellings included converted buses, trams and railway wagons as well as do-it-yourself villas in an infinite variety

139

of styles.   The results may be fascinating to the social
historian but to a geographer the lotissements and their
scatter of ill-assorted dwellings 'forment le plus souvent un
étonnant musée des horreurs' (Pinchemel, 1979, p.30).

By 1936 the population of the suburbs was greater than
that of Paris itself and there were already 36 communes with a
total of at least 25,000 inhabitants.   The lotissements had
spread well out into what are now the départements of the
Grande Couronne, this interwar vogue for individual housing
representing the second major phase in the suburbanization of
the capital (Figure 18).

After 1930 the rate of house-building declined with the
onset of the depression.   Forty-one thousand dwellings were
completed in 1930 in the département of Seine which included
Paris and its closer suburbs; by 1939 this total had fallen to
5,200.   Housing conditions worsened with the continued growth
of population and the loss of some of the existing stock to
office developments and later wartime damage.   It has been
estimated, for example, that the 85,000 dwellings completed in
the City of Paris between 1920 and 1945 were equivalent only
to the number lost to offices during those years (Bastié,
1964 b).   The situation continued to deteriorate for ten
years after the war when new building was minimal.   Only
55,000 new dwellings were completed in the whole of the Paris
agglomeration between 1945 and 1954 during which time the population increased by 620,000.   Some of these replaced houses
lost during the war and Bastié estimates that the new addition
to the housing stock was little more than 30,000 in those ten
years.   The slow pace of construction can be attributed partly
to the priority given in public investment to economic recovery,
but Bastié refers also to 'un état d'esprit antiparisien'
(1964, a, p.355) that was evident after 1947 following the publication of Gravier's Paris et le Désert Français.   Whatever
the causes of neglect, there was no doubting their consequences
when the abbé Pierre led his campaign against housing conditions in the capital during the winter of 1953-54 and when
the métro stations were left open at night as a refuge for the
homeless.

The years of neglect of Parisian housing were revealed by
the census of 1954.   No fewer than 60% of all dwellings in the
agglomeration of greater Paris had been built before 1914 and
this proportion rose to 85% in the City where the average age
of housing was 75 years.   Two-thirds of all dwellings in the
City of Paris had either one room or two and there was considerable overcrowding.   In the Bonne-Nouvelle quartier of
the 2nd arrondissement the gross density rose as high as
80,000 persons per sq.km.   Running water, cooking and sanitary
facilities were commonly shared in the crumbling apartment
blocks that accounted for the greater part of the housing stock.
Densities were lower in the suburbs on account of the inter-

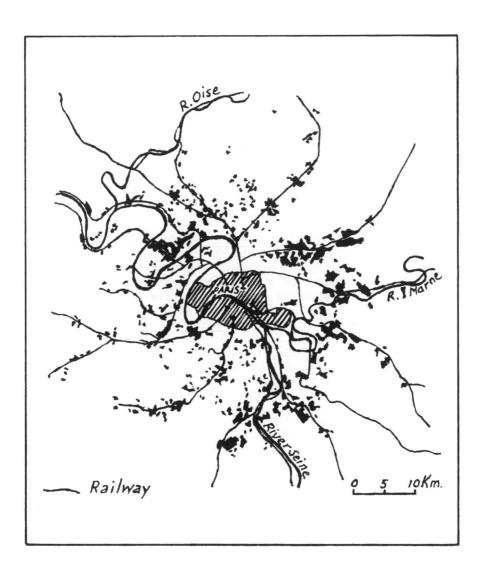

Figure 18.   Interwar lotissements of the Paris suburbs, after
             Bastié, 1964

141

war lotissements but, even here, over half the dwellings had
no more than two rooms and overcrowding was serious, especially
in the older industrial suburbs of the inner ring.

The response to the crisis was a housing drive that produced the grands ensembles. It was the start of a third major
phase in the evolution of suburban Paris. These grands
ensembles have already been described in Chapter 5 and will be
referred to only briefly here. Many of the dwellings were in
blocks of no more than four or five storeys, built so as to
avoid the expense of installing elevators, and examples can be
found of these stretching in a straight or slightly curving
manner for up to 300 metres. Their barrack-like facades would
be interrupted by occasional tower blocks rising to 12 or 15
storeys, sometimes more. Privately-developed versions would
often be described as 'residences'. Architecturally it was
an 'urbanisme de volumes', the ensembles grouping several thousand dwellings in their geometrically-arranged structures.
For the young families who occupied them they were little
better than 'machines à habiter'. The grand ensemble 'laisse
peu de place à la fantaisie et à l'initiative individuelle'
(Schéma Directeur, 1965, Avis et Rapport, p.299).

Sites chosen for the grands ensembles were frequently the
spaces left vacant by the speculators responsible for the earlier lotissements. In many cases these had been ignored
because they were considered unsuitable for housing, either
because of site difficulties, e.g. steep slopes, or because
they were environmentally unattractive on account of their
proximity to quarries and other nuisances. Some sites had
been neglected earlier because they were on plateaux, distant
from the railways which tended to follow the river valleys.
Government- or municipally-owned land was also favoured because
it avoided lengthy negotiations with private owners. Sometimes
it was land which had been used by the military or which had
earlier been set aside for public utilities or services.

Over half-a-million dwellings were completed in the Paris
Region between 1954 and 1962, many in the form of HLM apartments. But the population grew by more than a million over
the same period and the census of 1962 revealed housing conditions that were little better than those recorded eight years
earlier. In 1962 the mean size of dwelling in the Region was
2.58 rooms compared with a national figure of 3.09 rooms. The
Schéma Directeur (1965) noted that the 6 million rooms in the
Paris Region would have to be increased by 50% to 9 millions,
to achieve densities of occupation comparable with London or
New York. It proposed a construction rate in the future of
between 100,000 and 120,000 new dwellings a year.

This rate of new construction was reached for a time in
the late 1960s and early 1970s as the proposals of the Schéma
began to be implemented. The Housing Survey of 1978 states
that 105,000 new dwellings a year were completed in the Paris

Region between 1968 and 1975, a further 32,000 older dwellings being rehabilitated annually. Completion rates have fallen significantly, however, since 1973-74.

TABLE 10. AGE OF DWELLINGS

	City of Paris	Suburbs	Paris Region
Before 1949	75.7%	37.5%	48.7%
1949-1978	24.3%	62.5%	51.3%

Source: INSEE

Close to two-thirds of the housing stock of the Paris suburbs has been completed since the end of World War II (Table 10). The proportion is very much lower in the City of Paris, as one would expect - close to a quarter - although it exceeds a third in some arrondissements (13th, 15th) where there has been a large amount of urban renewal. Standards of comfort tend to reflect the age of dwellings and to be lower in the City. The fall in population and the high proportion of single-person households in the City (43.7% of the total in 1978) means, however, that overcrowding is often at its most severe, not here, but in the older suburbs of the Petite Couronne. Shortage of living space nevertheless remains a characteristic of the Parisian housing situation as a whole and for some families the problem is still acute. Katan (1981) quotes examples of immigrant families with several children occupying accommodation that offers no more than 20-30 sq.m of living space.

The problem of overcrowding in Paris and the inner industrial suburbs is compounded by a shortage of open space, above all of the kind in which children can play informally close to their homes. Paris has 2,200 hectares of public open space, equivalent to 9.5 sq.m per inhabitant, but the Bois de Boulogne and the Bois de Vincennes account for 1,840 ha. of this total, and they are peripheral to the City. Without them, the amount of open space per citizen falls to no more than 1.5 sq.m.

Employment and the Journey to Work

There were 4.6 million people employed in the Paris Region in 1975, equivalent to 22.6% of the active population of France as a whole. The dominance was greatest in the tertiary sector as one might expect, the Region accounting for 38% of those working in the country's offices. The proportion of women who are employed (40%) is significantly higher than the national average (30%).

The last 30 years have witnessed a movement in the Paris

Region away from manufacturing industry towards the tertiary sector, the latter accounting for 61.6% of all jobs in 1975 compared with 54.3% in 1962 (Table 11). The corresponding share of manufacturing has fallen from 36.1% to 29.5%. In part the fall in the importance of industry reflects the national trend towards the 'post-industrial' society, observable since 1962 (Chapter 2). It can be explained also by attempts on the part of the government to decentralize manufacturing from the capital although the contribution of these policies is difficult to measure and quickly becomes a matter of political debate. The total of manufacturing jobs lost to Ile-de-France between 1962 and 1975 - approximately 90,000 - hides, however, a considerable measure of redistribution which has been taking place at the same time within the Region and which has had far-reaching implications for the Paris agglomeration and its planning.

TABLE 11. EMPLOYMENT IN THE PARIS REGION, 1962-1975
(totals in thousands)

	1962 Nos.	%	1968 Nos.	%	1975 Nos.	%
Agriculture	66	1.6	53	1.2	40	0.9
Industry	1,445	36.1	1,399	32.7	1,356	29.5
Construction	318	8.0	368	8.7	368	8.0
Tertiary	2,177	54.3	2,452	57.4	2,838	61.6
Total	4,006	100.0	4,272	100.0	4,602	100.0

No more than a quarter of those who work in the City of Paris are now engaged in manufacturing whereas the proportion employed in industry in the suburbs is between a half and two-thirds. The contraction of manufacturing has been most severe in the City and the older suburbs, the rate of loss in the early 1970s being of the order of 33,000-35,000 jobs a year according to a special edition of Le Nouveau Journal (Vivre en Région Parisienne) published in May 1978. This study suggested that only 12% of the loss could be attributed to decentralization, however. Displacement of firms from crowded inner city sites to the outer suburbs was said to account for 42% of the loss, whilst 46% was attributed to firm closures due to financial failure, redevelopment schemes, etc. Enterprises suffering closure or displacement include both the specialist manufacture of the central city quartiers industriels and the heavier trades more typical of the older suburbs such as Pantin and La Courneuve to the north of the City where many thousands of jobs have been lost since the mid-1960s. Industries that have been most seriously affected include aircraft, electronics, machine-tools, clothing, printing and paper bag and carton manufacture.

Firms that have built new factories in the outer suburbs have shown a tendency to move in a radial direction from their old locations, maintaining on the new zones industrielles their linkages with other firms in the same industry. The same directional bias is evident in the moves that have taken place to towns in other parts of the Paris Basin or even beyond. This is well exemplified in the case of the Renault car factories which have followed the axis of the Seine downstream from their nucleus at Billancourt to Flins, Cléon-Elbeuf and Le Havre. The heavy engineering and chemical industries have been drawn to the north-west of the agglomeration whilst the electrical engineering trades have spread out in a westerly and south-westerly direction to Rambouillet,Dreux, Evreux and beyond, to Chartres, Nogent-le-Rotrou and even Angers and Cholet (Chapter 4). The pleasant living environment of the south-western suburbs has also proved attractive to the new science-based industries and their associated laboratories and research institutes. This 'intellectual axis' takes in the Centre d'Etudes Nucléaires at Saclay, university laboratories at Orsay and others belonging to the Centre National de la Recherche Scientifique at nearby Gif-sur-Yvette.

Despite the loosening out of industrial employment from inner Paris, there remains a wide mismatch between place of residence and place of work in the Region as a whole. It is a legacy of the traditional concentration of manufacturing in the west of the agglomeration and of the near total failure of employment to follow the untidy spread of suburbanization that characterized the interwar and early postwar years. It is a product also of the heavy concentration of office-based activity in central Paris and in one or two other nodes, the most important of which being that of La Défense. Pinchemel (1979) gives the total of those employed in the City of Paris as 1,885,000. Of this number, some 660,000 live in Paris itself, which means that 1,225,000 travel daily from the suburbs to work in the City.

Long journeys to work are a well-established part of the Parisian life style. The census of 1968 records an average journey time between home and work as one hour and twenty minutes. A typical journey might involve the use of the car or bus to reach the nearest suburban railway station, a 20-minute or half-hour ride on a crowded train, followed by a short trip on the métro and a walk to the office (Merlin, 1971). The use of more than one mode of transport was necessitated by the fact that the métro was built to serve little more than the City of Paris and by the frequent siting of grands ensembles at some distance from the nearest mainline station.

In addition to travel between the suburbs and Paris, the

pattern of commuting also involves a large number of journeys
between one suburb and another and also a reverse movement,
particularly of managers and executives, from the City to the
suburbs.   Thompson (1981) estimates that these latter categories
together account for as many journeys as the suburban to central
Paris flow.   Considerable congestion results from the amount
of cross-movement involved in these difficult kinds of work-
trip, the problem being compounded by the addition of journeys
to school, to the shops and for other purposes.   Pinchemel
(1979) suggests that the total number of déplacements involving
some form of transport (including the motor cycle and bicycle)
may be as high as 17 million each day in the Paris Region.
The use of the private car has increased and is particularly
important as a means of inter-suburban travel, reflecting the
inadequacies of the public transport system for this kind of
journey.   It has been encouraged also by extensions made to
the motorway system, bringing fast roads close to the heart of
the capital, though these advantages are offset in part by problems of traffic congestion and parking.   These latter problems
also affect the efficiency of the bus service which, in central
Paris, has a reputation for being slow, irregular and expensive.
       In view of the above it comes as no surprise that trans-
port issues should have received considerable attention in
successive plans for the capital.

Planning for Paris
       Congestion at the centre and confusion in the suburbs;
such, in a phrase, were the problems of interwar Paris.   In
the City, housing was old and overcrowded, there was little
open space, and existing buildings were ill-equipped to accommodate the expanding office-based industries.   Beyond the
City, the suburbs had grown in a random, unplanned manner and
were singularly ill-provided with even basic services.   Public transport was inadequate, especially for the increasingly
long journeys that were undertaken.   Perception of the problems tended to be obscured, however, by the prosperity of Paris
relative to that of the rest of France, whilst the fragmented
nature of local government outside the City acted as a deter-
rent to the search for solutions involving the whole of the
agglomeration.
       One of the first to recognize the need to plan, not only
for Paris itself, but also for a wider urban region, was
Henri Sellier who served as councillor for the département of
Seine and later as minister in the Poincaré government of the
late 1920s.   Largely as a result of his pressure a committee
was set up in 1928, the Comité Supérieur de l'Aménagement et
de l'Organisation de la Région Parisienne, charged with the
responsibility of producing a plan for such a region.   The
area defined for study extended for some 35 km from Notre Dame,

taking in the whole of the département of Seine and certain communes from the départements of Oise and Seine-et-Oise, a total area of 3,800 sq.km. A plan was produced in 1934 and submitted to the 657 communes involved but the resultant delays were so protracted that the government felt obliged to over-ride opposition and grant its formal approval to the plan in 1939. By this time, however, the recommendations of the plan - usually known as the Prost Plan from the name of one of its authors - were already out of date and the war intervened to halt their application.

A weakness of early planning was its preoccupation with the problems of the present and its failure to look ahead and make proposals for anticipated future developments (Carmona, 1975). The Prost Plan had sought to control the spread of lotissements and to provide better services but the many years involved in its formulation meant that little was achieved.

In 1941 a Commissariat aux Travaux de la Région Parisienne was set up and at the same time the planning region was rede-fined to take in the whole of three départements (Seine, Seine-et-Oise, Seine-et-Marne) as well as five cantons from the département of Oise. It is of interest as a forerunner of the present Paris Region. Subsequently, in 1943, a Service Tech-nique d'Aménagement de la Région Parisienne (SARP) was created with the task of preparing a new plan. It was to be advised by a reorganized successor to the 1928 Comité d'Aménagement de la Région Parisienne (CARP). Again there were long delays, caused principally by the need to attend to postwar reconstruc-tion, and the plan produced by SARP was not submitted to the government for scrutiny until 1956. Meanwhile the office of Commissaire à la Construction et à l'Urbanisme pour la Région Parisienne had been set up in 1955. The new commissioner's duties were not fully defined until 1958, however, when he was given the responsibility of producing, with the assistance of SARP, yet another plan for the region. This was the Plan d'Aménagement et d'Organisation Générale de la Région Parisienne (PADOG) which would take account of the proposals that still seemed relevant in the plan put forward two years earlier in 1956.

The PADOG proposals were published in 1960, an admirable improvement on the time required to produce earlier plans, and an advance also in that the recommendations looked forward, anticipating the changes that were expected to take place during the ten years between 1960 and 1970. The plan took account of the major new housing developments, mainly in the form of grands ensembles, that were taking place at Massy-Antony, Créteil, Saint-Denis and elsewhere, and proposed a number of functional nodes that would introduce a degree of order to these burgeoning suburbs. It was suggested that they should be at La Défense, on the plateau of Vélizy-Villacoublay in the southern suburbs, at Le Bourget, possibly in connection

with the transfer of the airport to a new site further north,
and at a couple of unspecified sites to the south-east and east.
Attention was also given to the need to improve public transport services and to make provision for the rapidly growing
number of private cars.   An express rail service was proposed
with a west-east axis running from La Défense to Boissy-Saint-
Léger and a north-south one from the Gare du Nord to Versailles.
Plans for new roads included the Boulevard Périphérique and
other autoroutes serving the suburbs.

Several of the recommendations in the PADOG have since
been carried out, but the plan itself soon attracted criticism
for its overall emphasis on restraint.   It was a product of
the years when national planning was preoccupied with the
decentralization of Paris and it reflects the efforts then
being made to direct industry and other activities to the
provinces.   Underlying the plan was a belief that the growth
of Paris could be controlled and that the population of the
capital would not exceed 9 millions.   In support of this policy
of restraint, the limits of the agglomeration were defined by
means of a firm boundary beyond which was a 'country' zone
where development was to be severely restricted except in
'satellite towns', such as Mantes, Melun and Fontainebleau.
Parallels are suggested with the role played by London's Green
Belt.   But the pressures were too great to be contained in
this manner, as many as 20,000 dwellings being erected outside
the permitted zone within four years of the plan's appearance.
With the Region's population growing by more than 150,000 a
year, a policy based on physical containment was soon seen to
be unrealistic.   Within a few years it was to be abandoned in
favour of a policy that sought to accommodate growth.

Meanwhile, in order to implement the proposals of PADOG,
it was necessary to ensure the cooperation of all the local
authorities involved.   To this end the District de la Région
de Paris was established in 1961.   Covering the same three
départements as the earlier planning region, this District was
not a unit of local government comparable with, for example,
the Greater London Council, but had the status of an établissement public with access to funds in order to carry out studies
into the planning needs of the Region and to assist in their
realization.   The first head (Délégué Général) of this new
public body was Paul Delouvrier and one of the earliest studies
undertaken was into the likely future growth of population in
the Region.   The results of this enquiry, published as part
of a livre blanc in 1963, suggested that the Paris Region would
probably house 11.6 million people in 1985 and 14 millions by
the end of the century.   It is ironical that this publication,
which more than anything else put an end to the containment
philosophy of PADOG, should have come from a body created
initially to advance the proposals included in that same plan.

The Schéma Directeur

Once the notion of containment had been rejected, it became necessary to prepare a new plan, and the responsibility for this was entrusted by M.Delouvrier to the Institut d'Aménagement et d'Urbanisme de la Région Parisienne (IAURP) which had been established in 1960.   Work on the Schéma Directeur was carried out during the course of 1963 and 1964. It proceeded in some secrecy in order to avoid speculation in land at sites where proposals for development were being made and, in fact, maps with conflicting information were actually prepared in order to mislead those who were seeking to profit from the intentions contained in the plan.   In addition, some 45,000 ha. of land were made subject to the ZAD procedure. The plan was revealed to government ministers towards the end of 1964 and to the public in June 1965.

The Schéma Directeur was based on two fundamental assumptions, the first being that the Region would experience population growth on a large scale, and the second being that the space needs of the population would be greatly enlarged as living standards rose.   The population estimates on which the Schéma was based are set out in Table 12 and it is of interest to note that, although the Region was expected to house 14 million inhabitants by the end of the century, this would be a lower proportion of the national urban population than in 1962. The Schéma was a plan to accommodate growth but not at the expense of other parts of the country and of strategies already formulated for promoting development elsewhere.   It was anticipated that the population of France as a whole would reach 60 millions by 1985 and 75 millions by the end of the century.

TABLE 12.   ANTICIPATED GROWTH OF THE URBAN POPULATION (millions)

	Urban Population of France	Paris Region	Rest of France	Percentage in Paris Region
1962	29.5	8.4	21.1	29%
1985	44.0	11.6	32.4	26%
2000	58.0	14.0	44.0	24%

Source: Schéma Directeur

The working population of the Paris Region, approximately 4 millions in 1962, was expected to grow to 6.2 millions by 1985.   More than two-thirds of this increase (1.6 million jobs) would be accounted for by growth in the service sector which, it was thought, would almost double the size of its workforce. Expansion of the industrial labourforce was expected to be of more modest proportions, the 600,000 new jobs representing a

30% increase on the 1962 total.

The authors of the Schéma laid stress on the need to provide more generously for the accommodation of the existing population as well as for the anticipated increase. It was suggested that, allowing for garden and car parking, a resident of Paris in the mid-1960s could reasonably expect an allowance of 100 sq.m of living space compared with a prewar 'norm' of 35 sq.m. Given that the number of dwellings would very nearly double by the end of the century (from 3.2 to 6 millions) and that the number of rooms would have to increase three-fold to reduce overcrowding, there would need to be a much more generous allocation of land for housing purposes. The proposals in the Schéma were, in fact, based on the belief that there should be a four-fold increase in the amount of land devoted to residential use. New and rebuilt factories and offices would also require more space and the Schéma anticipated a doubling of the amount of land devoted to industry and a three-fold increase in that given over to offices.

Other forecasts made in the Schéma were of a growth in purchasing power of the order of three to four times by the close of the century, of a progressive increase in the amount of leisure time, and of an expansion in private vehicle ownership from 1.7 millions to 5 millions. The number of journeys undertaken each day by all modes of transport was also expected to increase by between three and four times. It was clear that such needs could be satisfied only by making far-reaching changes to the transport network and to the provision of shopping, recreational and other services.

A number of alternative strategies were considered for accommodating the expected growth. These included a plan for a single new city (centre-bis or Paris-Parallèle) outside Paris, rejected because of its inflexibility, and a proposal to build a dozen or so new towns on the London model outside a restrictive green belt (Figure 19). The latter did not commend itself either. It was felt that such a plan would perpetuate the concentric pattern of growth and, by continuing to stress the functional role of the core, would add to traffic congestion and associated problems. The preferred solution was for growth axes that would reduce pressure on the radial routeways and for a limited number of new centres large enough to attract employment and to offer services of an order sufficient to offset some of the dominance of central Paris. It was a strategy that has been described as one of 'déconcentration concentrée' (Cohen, 1978). The question remained of the direction to be taken by these axes of growth. One possibility was to follow the important routeway afforded by the Oise valley towards the industrial Nord/Pas-de-Calais; another was to utilize the valley of the Marne for an extension to the east. Both were rejected in favour of two preferential axes following the general direction of the Seine valley from south-east to north-

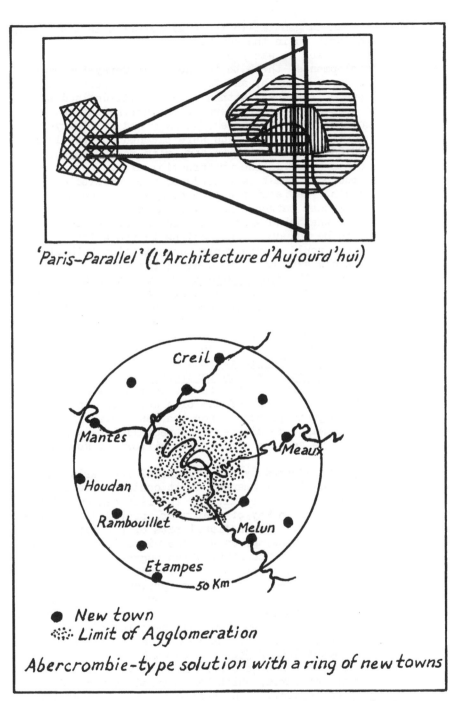

'Paris–Parallel' (L'Architecture d'Aujourd'hui)

New town
Limit of Agglomeration

Abercrombie-type solution with a ring of new towns

Figure 19.   Two unadopted strategies for the Paris Region

west, one of them tangential to the agglomeration in the north, the other similarly so in the south.

Along these axes it was proposed to carry out eight major new urban developments, three to the north at Cergy-Pontoise, Beauchamp and Bry-sur-Marne, and five on the southern axis. The site of one of these was within a rather ill-defined area to the south of Mantes; others were to the north-west and south-east respectively of Trappes, at Evry and at Tigéry-Lieusaint. Most of the new towns were planned for populations of between 500,000 and one million and, with redevelopments proposed at other nodes on the preferential axes, they were expected to absorb at least two-thirds of the anticipated population growth. Attention was to be given to the preservation of open spaces between the new urban centres and also in the valley of the Seine where the demands for development were expected to be reduced. Extensions to the existing transport network would ensure rapid communication both along the line of the new axes and with central Paris.

Publication of the Schéma Directeur aroused great interest, not only in Paris, but in other parts of France where it did much to promote the merits of the broad-brush (or structure plan) approach to urban planning and to prepare the way for the legislation of 1967 (Chapter 5). But it also drew adverse comment. It was argued that the new towns were too peripheral to serve as 'pôles restructurants' and that the role of other places designated 'centres restructurateurs' in the suburbs was ill-defined and left too much to the initiative of the private sector. The authors of the plan were also accused of failing to integrate proposals for the new towns with those relating to other major developments, especially the routes chosen for motorways and the site of the airport at Roissy-en-France.

The Schéma was adopted by the council of the Paris District in 1966 but calls for changes to it continued to be heard. Publication of the results of the 1968 census which followed the upheavals in May of that year, provided an opportunity of taking stock. The national rate of population growth was seen to be slowing, as was the migrational pull of Paris over the rest of the country, which suggested that the forecast of 14 million inhabitants by the end of the century was unlikely to be reached. But if the pressure for new development was thereby reduced, the problem of internal disequilibrium had, if anything, become worse. A high proportion of the new jobs created in the 1960s had been in offices in central Paris which, because of the reduction in the resident population, had meant longer journeys to work. What is more, three-quarters of these new jobs were located west of a line through the centre of Paris, which half already possessed 62% of the Region's employment. The west/east contrast had thus been exaggerated. Meanwhile the number of vehicles on the

roads of greater Paris had risen from 850,000 in 1954 to some
2.4 million by 1968 whilst the public transport services were
faced with a problem of mounting losses.

Changes affecting the administration and planning of the
Paris Region also made it appropriate to re-examine the Schéma.
Legislation introduced in July, 1964 made provision for the
abolition (except for Seine-et-Marne) of the départements which
made up the Region and their replacement by the present eight
départements.   This became fully effective from the First of
January, 1968, the administrative needs of the new départements providing opportunities for growth in the service sector,
especially in those places selected to serve as préfectures.
The regional reforms, implemented in 1966, were important in
strengthening the planning powers of the Region through the
executive role of the regional préfet.

Amongst the critics of the 1965 Schéma was M.Chalandon,
appointed Minister of Equipment in 1968.   Paul Delouvrier,
with whose ideas he had little sympathy, was soon succeeded
as Délégué (and Préfet de Région) by Maurice Doublet and it
was to satisfy government criticisms as well as to take
account of the changed circumstances that a revision of the
Schéma was undertaken during the course of 1969.

The 1969 revision resulted in the number of new towns
being reduced from eight to five; their projected population
totals were also lowered to within the range 300,000 to
500,000.   The proposed development south of Mantes was abandoned altogether in order not to prejudice the plans for a
new town at Le Vaudreuil in the lower Seine valley.   Elsewhere modifications were made in response to local objections,
whilst also reflecting the desire to strengthen the employment potential of the eastern suburbs.   Thus the two proposals for Trappes were conflated in a single new town of
Saint-Quentin-en-Yvelines which was sufficiently distant from
Versailles not to harm the latter.   Conversely, the site of
Tigéry-Lieusaint was displaced southwards so as to avoid dup-
lication of the facilities of Melun and was renamed Melun-
Sénart.   A desire not to prejudice the success of Evry was
also a factor in the change of site.   To the north-west of
Paris the proposed new town of Beauchamp was also abandoned
as too close to Cergy-Pontoise and to preserve the wooded
countryside there.   The developments at Cergy-Pontoise, Evry
and in the Marne valley were confirmed, the latter taking the
name of Marne-la-Vallée.   Of the five new towns originally
proposed for the western suburbs, only two remained, whilst
all three were retained in the east.   The suggestion had
been made that an additional new town site be designated to
the north of the new Paris-Nord airport in association with
that development, but the idea was not adopted in the 1969
revision of the Schéma.

Plans were retained in the Schéma to restore and reinvigorate a number of centres in the older, inner suburbs including Bobigny, Créteil and Nanterre which had been selected as préfectures of their respective départements. Proposals were also made for Le Bourget where the existing airport was to be run down and for Vélizy-Villacoublay in the southern suburbs, bringing the number of such major redevelopment schemes to seven. Support for public transport was reaffirmed in the 1969 revision and various suggestions made for extensions to the RER (below) and métro system. Finally, the influence of the Minister may be seen in the decision to permit low density urbanization, mainly of individual dwellings, in three areas of the outer suburbs: in the valley of the Mauldre to the west, towards the new airport in the north, and in the valley of the Révillon to the south-east. It was argued in justification of the new proposals that there was spare traffic capacity on existing or planned motorways and express rail routes, but critics have found it hard to reconcile these additions on the urban fringe with the new towns strategy and also with the desire to sustain a viable public transport service, Pinchemel (1979) for example, describing them as essentially contradictory and a return to the radial-concentric concept of city growth. Others, however, have pointed out that private developers were already promoting schemes outside the context of the Schéma and that it was more realistic to try and direct this growth than to oppose it.

The Schéma Directeur of the Paris Region did not receive formal government approval until July, 1976. The long delay in granting such approval did not prevent work being undertaken on many of the proposals set out both in the Schéma and in the earlier Plan d'Aménagement of 1960, but it also left the door open to further revisions and these were, in fact, undertaken in 1975. The presidential election of 1974, followed by the publication of the results of the 1975 census, were circumstances which encouraged such a review. Anxiety was also being felt over the consequences of the recent huge increase in crude oil prices.

The census revealed a further slowing in the rate of population growth and by 1975 the projected total living in the Paris Region at the end of the century had been reduced to 12 millions. Other facts noted were the shrinkage of manufacturing and the importance of the service sector with over a million and a half persons now working in offices in the Region. The oil crisis cast doubts on the wisdom of urban motorway construction and strengthened the argument of those calling for greater investment in public transport. This point of view was echoed in growing support for the ecology movement and in the much greater interest now being shown in the landscape and in matters of conservation.

Social and environmental issues are much more evident in

154

the 1975 revision than in earlier editions of the Schéma. There is a call for better hospital, higher education and shopping provision in the suburbs; lengthening journeys to work are deplored, and concern is expressed over the problems resulting from inner city decay. The five new towns are retained as foci of employment and higher order services in the outer suburbs but their planned population totals are cut back to 200,000 in accordance with the more modest expectations of population increase in the Region as a whole. Some opponents had wished to see the number of new towns reduced to three. Greater investment in the tertiary sector is also recommended for the seven pôles restructurateurs in order to help them in their task of improving the 'quality of life' in the most densely populated parts of the agglomeration. The motorway building programme is reduced by cutting out several sections of radial that would have converged on the Boulevard Périphérique and correspondingly greater attention is paid to public transport, especially rail services.

There are more positive proposals for protecting the countryside in the 1975 revision than in earlier versions of the Schéma. In particular the plan designates five zones naturelles d'équilibre (ZNE), areas that are still predominantly rural in appearance and land use where the policy will be to preserve the landscape and protect the interests of farmers whilst at the same time making provision for the growing recreational demands of the population. Together the five occupy a fifth of the total area of the Paris Region and amount almost to a green belt (Figure 20). Other proposals are made for woodland management and tree-planting.

The change between 1965 and 1975 from a quantitative approach to planning to one where the stress is much more qualitative in its outlook is apparent from the above. The Schéma received its formal approval on 1 July, 1976, the date which also brought into operation a number of changes in the organization of the Paris Region, now to be known as Ile-de-France. As a result of these, the Region acquires planning powers similar to those of the 21 other French planning regions. The Regional Council has not been slow to exercise its authority and already by 1978 a number of revisions to the Schéma were being requested. As a result of these, the direction of planning for the Region is even more towards welfare, environment and 'quality of life'. Ways of satisfying the desire for individual houses are being discussed, whilst the target populations of the new towns have been still further lowered to between 100,000 and 200,000. Planning by the early 1980s had become more humble, and more humane.

Change at the Centre
The Schéma Directeur of 1965 was a regional structure plan, concerned more with the suburbs and their surroundings

Plaine de France

Plaine de Versailles

Hurepoix

Plateau de Brie

Plateaux du Sud

Principal built-up area

New Town

0   10 Km.

----- Zone naturelle d'équilibre

NEW TOWNS

CP Cergy-Pontoise
SQ Saint-Quentin-en-Yvelines
MLV Marne-la-Vallée
E Evry
MS Melun-Sénart

SUBURBAN NODES

LD La Défense
SD Saint-Denis
B Bobigny
RO Rosny
C Créteil
RU Rungis
V Vélizy-Villacoublay

Figure 20.   1975 revision of the Schéma Directeur for the
Paris Region, after Pinchemel, 1979

156

than with the central city, the Ville de Paris.   The City, in
fact, published its own Schéma in 1967 which was intended to
complement the regional document.
     This City plan, on which work had been proceeding ever
since 1959, made various proposals for the historic core where
restoration was to be a major theme, for a number of 'pôles'
corresponding with the principal railway stations where development of offices was to be encouraged, and for the peripheral
arrondissements where there was to be extensive urban renewal
and replacement of factories by employment in the tertiary
sector.   Extensions to the underground rail network were
recommended but the most striking feature of the transport proposals was the emphasis which it placed on motorways, a network of expressways being suggested which included a north-
south axial route and another along the left bank of the Seine.
     The plan was attacked for the invitation it afforded to
speculative development and for the destruction that would
accompany the building of the motorways.   It was never formally adopted by the City Council and a revised Schéma was
published in 1976.   The latter, influenced no doubt by the
City's falling population, places more stress on the need for
new housing.   Less land is set aside for office use and the
motorway programme is reduced in favour of public transport.
     Both the above plans set out to present objectives in
broad terms.   They have had some influence in guiding the
pattern of development but the progress of re-investment and
renewal has not been dependent upon them and indeed the face
of Paris had been substantially transformed before the 1976
Schéma appeared.   In practice, changes have been guided by
the normal land use controls of ZUP, ZAC and ZAD, as well as
by a series of local schémas and plans.   Initiating the
developments have been a host of organizations which include
government departments, HLM agencies and a wide range of private and semi-private (mixed economy) companies.   A report
prepared for the City of Paris by a research team headed by
Professor Jean Bastié, and published in 1975 under the title
of Vingt Ans de Transformation de Paris: 1954-1974, describes
the extent of change that has taken place in the City over
this period.
     The report chronicles the urban renewal that has affected
no less than a quarter of the entire area of the City, a transformation comparable in its scale with that undertaken by
Baron Haussmann between 1850 and 1870.   Road building and
improvement schemes have added over 20% to the amount of land
devoted to the movement of motor vehicles, but the most striking change has been the increase in total floorspace, both
residential and non-residential, as a result of the construction
of high buildings.   Some 8.8 million sq.m of floorspace were
demolished between 1954 and 1974, according to Professor Bastié's
research team, but this was replaced over the same period by

23.5 million sq.m of new floorspace, a net gain of 14.7 million sq.m.   The amount of land devoted to industry and warehousing has shrunk by more than a quarter and the increase is attributable to the growth in office employment and to the additional floorspace devoted to public services and housing.

New housing accounts for no less than 16 million of the 23.5 million sq.m of floorspace built since 1954, a surprisingly high proportion perhaps in view of the shrinkage of the City's population by some 600,000 over the same period.   The number of dwellings has grown by more than 200,000 over the 20 years in question and Bastié accounts for the apparent anomaly of a declining population and an enlarged housing stock in terms of social trends, particularly those in the direction of vieillissement and embourgeoisement.   There are more old people and more elderly couples or single persons living in apartments that would formerly have housed families with children.   The same problem has been examined by Gottmann (1976) who attributes the need for more housing to the size of the City's 'floating' population of transients and 'transhumants'. Gottmann notes that a large proportion of the new dwellings have been expensive apartments, little more than a fifth of the total being 'social housing' of the kind built by the HLM agencies.   He suggests that many of these apartments are likely to be, either the pieds à terre of the national and international business and diplomatic community, or the second homes of bourgeois residents who prefer for tax and other reasons to count their country house as their primary residence and maintain their town flat as résidence secondaire.   These are the 'transhumants', typical of the modern city and its contemporary life styles, whose presence, together with that of the transients, means that the actual resident population of Paris is likely to be higher than that recorded by the census.

Renewal and redevelopment has been a source of considerable controversy in Paris.   Three themes are recurrent: loss of the old vie du quartier, the treatment of historic buildings and cherished sites, and the intrusion into the famous skyline of modern tower blocks.   Restoration in the district of Le Marais has drawn criticism both for its effect on the pre-existing community and its treatment of some of the fine hôtels (Chapter 7), but the fiercest argument has centred on the treatment of Les Halles.   The dispute has led to modification of the plans for this former market site, more land being reserved as open space, but not before the destruction of the old food halls had taken place.

Major schemes of redevelopment involving tall buildings have taken place at Fronts-de-Seine, Maine-Montparnasse, Porte d'Italie, Gare de Lyon and elsewhere.   Of these the most controversial has been that of Montparnasse where, from amongst a mass of slab-like structures, a huge tower rises more than

200 metres.   Ardagh (1980) describes it as a disastrous aber-
ration, and Cohen (1978) as 'le nadir de l'architecture
française'.   The latter also reserves harsh words for the
faculty of science built on the site of the old wine market
by the Seine which he condemns as 'une des dix pires
constructions du monde moderne'.   Most of these unattractive
structures were erected in the late 1960s and early 1970s when
the prevailing philosophy was one of expansion, profit and
prestige.   Since 1974 the controls on redevelopment have been
tightened and M.Chirac, who took up the newly recreated office
of mayor in 1977, has tried to reconcile conservation with
renewal.   The lack of an agreed overall plan meant that many
of the earlier developments took place in a piecemeal manner,
the result too often being a lack of harmony between new and
old and of coherence between one new scheme and another.
This fragmented kind of development is less likely in the
future which will probably see greater attention paid to the
continuing imbalance between west and east, with increased
investment in the new employment nodes at La Villette in the
north-east of the City and at Austerlitz-Bercy in the south-
east.

Transport Planning
     Despite the continuing problems of traffic congestion
and parking, the improvements that have taken place in the
public transport network are widely recognized as one of the
success stories of postwar planning for Paris.   Amongst these
the creation of the Réseau Express Régional (RER) rail service
to complement the older main line and métropolitain systems
must be acknowledged as the greatest single achievement.
     A longstanding rivalry between the main line railway com-
pany (SNCF - Société Nationale des Chemins de Fer Français)
and the Paris public transport authority (RATP - Régie
Autonome des Transports Parisiens) has been one of the principal reasons for the length and slowness of journeys to work
in the French capital.   Uncertainty over the role of the
métro delayed its opening until 1900 and the choice of a different gauge from that used by the main line companies meant
that it would never provide the kind of link between main
line stations which the government had wanted.   One of the
aims behind the building of the RER, which has the same gauge
as the SNCF, has been to provide this link, and indeed land
which had once been reserved for lines joining the main line
stations has been used in the construction of the RER.
Another object of the RER has been to join City and suburb in
a way that the métro, which had few extensions beyond the
City's boundary, was never capable of doing.   The RER also
has a greater traffic capacity than the métro and is faster
on account of the wider spacing of its stations (which averages
only 500 m on the métro).

The intention to build a réseau express is usually associated with the plan to create a new centre d'affaires at La
Défense, formulated in the mid- to late-1950s.   Work began
on the first section of the RER in 1961 and it featured prominently in the Schéma Directeur of 1965 which showed a network
joining the new towns (except Mantes) to each other and to
the centre of Paris.   For the most part the system was to be
underground within the City but a surface railway in the
suburbs.   Construction difficulties delayed the completion of
this first section until 1969.   Thereafter new stretches of
line were added but it was 1977 before the RER took on the
appearance of a network.   In December of that year the final
gap was closed in the 55 km line running from Saint-Germain-
en-Laye in the western suburbs through the centre of Paris to
Boissy-Saint-Léger in the south-east.   From the new station
of Châtelet-les-Halles, which claims to be the largest and
busiest underground station in the world, a second line ran
as far as Saint-Rémy-lès-Chevreuse in the south-western
suburbs ('ligne de Sceaux').

Since 1977 there have been further additions to the network and, in order to reduce costs, modifications have been
made to the 1965 plan which involve making better use of the
existing main line rail system (Figure 21).   In particular,
interconnections have been established at several stations
which permit direct running by trains over both the RER and
SNCF lines.   A new underground station was built at Gare du
Nord, for example, which links the RER system with the suburban
line serving Charles de Gaulle airport.   At Gare de Lyon
there is a similar interconnection of the RER with lines to
the south-east which serve, amongst other places, the new towns
of Evry and Melun-Sénart.   From Nanterre on the RER network,
the SNCF is building a new line to Cergy-Pontoise.   Finally,
the SNCF has created a new express service across Paris on the
left bank by joining its stations at Gare d'Orsay and Gare des
Invalides, thereby permitting through running from Versailles
and Saint-Quentin-en-Yvelines in the south-western suburbs
through the centre of the City to Orly airport and, beyond, to
Dourdon and Etampes.

The RER connects at many points with the métro as well as
with the SNCF, and a number of extensions of the métro into
the inner suburbs have also been made.   The object of linking
the new towns and other suburban nodes by fast rail service to
the centre of Paris has been achieved, but less attention has
been paid to inter-suburban connections than was envisaged in
the 1965 plan.   Travel between one suburb and another still
depends on the bus or, more likely, the private car which has
resulted in heavy vehicular use of the Boulevard Périphérique
as a link between the various radial expressways.   Parts of
a super-périphérique, the A.86, have been built but this outer
circular is incomplete and plans to extend it through the

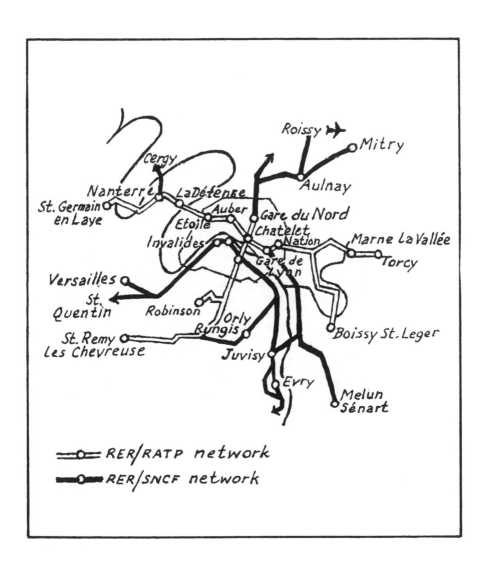

Figure 21. The Réseau Express Régional

western suburbs have aroused considerable opposition on environmental grounds. An alternative suggestion has been that the
'grande ceinture' rail line be reopened for passenger services
as an addition to the RER network.

Growth Nodes in the Suburbs
    Successive plans for the Paris Region have recognized the
need to restructure the overcrowded inner suburbs. This was
to be done, not only by building new housing, but also by the
introduction of new forms of employment, especially in the
tertiary sector. In order to attract higher order services,
however, it was considered necessary to concentrate investment
on a limited number of centres offering economies of scale and
the advantages of a lively contact environment. Seven have
been developed: La Défense, Saint-Denis, Bobigny, Rosny, Créteil,
Rungis and Vélizy-Villacoublay (Figure 20).
    La Défense, on the peninsula of land formed by the first
big bend of the Seine below Paris, was where the final stand
against the Prussian army took place in 1870. In the late
1950s it was a typical untidy suburb of the inner ring with
narrow streets, small factories and poor housing. The
decision to redevelop it as a centre restructurateur de
banlieue was taken in 1955 when it was thought that by creating
a new centre, particularly for offices, it would also be
possible to take some of the pressure off central Paris where
the loss of housing to offices and the growth of commuting was
giving rise to serious concern. Lying astride the continuation
of the Champs Elysée in a westward direction, La Défense would
serve as an extension of the Parisian central business district.
    A development board, the Etablissement Public d'Aménage-
ment de la Défense, was set up in 1958 to oversee the planning
and creation of the new centre. Loans to the EPA were made
by the Caisse des Dépôts et Consignations and by the early
1980s nearly a million of the projected 1.5 million sq.m of
office floorspace had been constructed. The office towers
flank a central spine or esplanade which is reserved for pedes-
trians, this deck being built above the road and rail routes
and the associated carparks and stations which serve the complex. A major shopping precinct was completed in 1980 which
has attracted representatives of the grands magasins, La
Samaritaine and Le Printemps, as well as a wide range of luxury
shops and boutiques. With 110,000 sq.m of shopping floorspace
it is the largest commercial centre of its kind in France.
In addition, there are restaurant and entertainment facilities,
including a striking Palais des Expositions. Accommodation
in the form of apartment blocks is also being built for up to
20,000 residents, some of whom will occupy 'social' housing.
    The development of La Défense is now well advanced and
some of the original aims of the project are being realised
as major companies - IBM, Citibank, Rhône-Poulenc,

Saint-Gobain - move their headquarters here. The role of this office and commercial centre as a 'restructuring node' has been complemented by the creation on a nearby site ('Zone B') of the préfecture of Hauts-de-Seine and the university campus of Nanterre, together with parks and public gardens.

The most distinctive of the other centres restructurateurs is Rungis which focusses on the huge food market opened here in March, 1969. The formal decision to move the market from Les Halles was taken in 1962 and work began on the Rungis site two years later directed by a mixed economy company, La Semmaris (Société d'Economie Mixte d'Aménagement et de Gestion de Paris-Rungis). Rungis is by far the largest of the nineteen marchés d'intérêt national (MIN) which the government had earlier agreed (1953) to set up in order to modernize the marketing and distribution of foodstuffs in France. The site of the market, 11 km south of the centre of Paris, is adjacent to the Autoroute du Sud and Orly airport and well-served by rail. Closely associated with it is the regional shopping centre of La Belle Epine opened in 1971 with 86,000 sq.m of floorspace, and the 'Zone du Delta' which offers hotel accommodation, restaurants, cinemas and a congress hall (Brayne, 1972).

Of the remaining nodes, Créteil has been described as a new town in all but name (Steinberg, 1976). Early development took the form of a ZUP, designated in 1959, and the building of a grand ensemble of 6,000 dwellings at Mont-Mesly. These were typical of what were being erected elsewhere in the Paris suburbs and it was a desire to achieve something better that led Créteil's energetic and gaullist mayor, General Billotte, elected in 1965, to set up a mixed economy company in order to carry out an ambitious programme of planned growth. The métro was extended to Créteil in 1968, and in the same year a Carrefour hypermarket opened, marking the first stage in the creation of a regional shopping centre.

The new development at Créteil has attracted much attention for the quality of its landscaping. Apartment and office blocks have been built in a parkland setting, the hope being that many of the residents will be able to walk to work rather than undertaking the tiring journeys that have been the lot of most suburban dwellers. With 250,000 sq.m of office space, Créteil is second only to La Défense amongst the new suburban centres, its attraction to higher order services being enhanced by the establishment at its core of the préfecture of Val-de-Marne. A university campus is also included in the redevelopment.

With Bobigny and Rosny, Créteil is also in the eastern suburbs, one intention behind the promotion of these centres being that they should help to overcome the longstanding imbalance between west and east in Paris. In this respect they are also complementary to the new towns.

163

Chapter 9.

NEW TOWNS

New towns have been planned and built in many countries
in order to resolve postwar housing shortages or as part of a
strategy of regional and resource development.   There are
more than 30 in Great Britain and many hundreds in the Soviet
Union where new towns are closely associated with the spread
of industrialization.   They have been built in the United
States as part of the 'new communities' programme introduced
in the late 1960s, and in less developed countries in an
attempt to offset the primacy of the national capital (Merlin,
1969).
     In contrast with many of these other countries, France
has made limited use of the new town formula, preferring to
find solutions to housing and other urban problems within the
framework of existing administrative structures rather than
setting up new institutions that might challenge the jealously-
preserved roles of ministry, département or commune.   Only
nine places enjoy formal designation as new towns, five of
which are in the Paris Region, having their origin in the proposals of the Schéma Directeur of 1965 (Chapter 8).   The
remaining four are associated with plans for provincial cities
drawn up by metropolitan planning authorities (OREAM) or, in
the case of Le Vaudreuil, with plans for the lower Seine valley
(Chapter 3).   Other cities have planned major extensions to
their urbanized area - Toulouse-Le-Mirail, Villeneuve-
Eschirolles, Herouville Saint-Clair, etc. - and it has been
suggested that these could qualify as new towns in a broad
sense, but none of them are directed or financed in the distinctive manner of the nine (Rubenstein, 1978).
     Following publication of the regional plan for Paris in
1965, work on the new towns began the following year when
legislation was introduced which made it possible to appoint
a directeur responsible for directing a mission d'études et
d'aménagement.   This mission is made up of a team of 30 to
40 specialists who, together, are responsible for establishing
a master plan for the new town.   The work of the several

missions was coordinated by a Groupe de Travail Interministériel which, as its title suggests, included representatives
from the various ministries involved (principally Finance,
Interior and Equipment) together with préfets and the directeurs
concerned.   In the early years the Groupe was presided over
by the Delegate General of the Paris Region, but after the with-
drawal of M.Delouvrier from the scene in the late 1960s, the
role of the regional préfecture was reduced and the organization became known as the Groupe Central des Villes Nouvelles
(Brissy, 1974).   By coordinating the investment programmes of
the various government departments involved it helps to ensure
the balanced development of individual new towns.

One of the principal difficulties facing the missions in
their work of creating a master plan was the autonomy of the
individual communes involved.   The nine new towns as a whole
extend over more than a hundred separate communes and, in
France, the communes have traditionally resisted suggestions
that they be fused into larger units of local government.   The
syndicat intercommunal à vocation multiple was introduced in
1959 in order to facilitate cooperation between communes willing to share some of their responsibilities, and this formula
was used in the early stages of planning at Evry where it was
necessary to involve only four communes.   After 1966 it was
also possible to establish a communauté urbaine, but this was
more suitable for the planning of existing urban areas than
for the creation of new ones.

It was to overcome the problem of collaboration amongst
existing units of local government that the 'Loi Boscher' was
adopted in 1970.   Michel Boscher was a député and, at that
time, deputy mayor of Evry, and his legislation made provision,
first for the delimitation of a périmètre d'urbanisation which
might or might not coincide with existing local government
boundaries and, secondly, for the establishment of a syndicat
communautaire d'aménagement (SCA) to ensure cooperation between
the communes concerned.   Without the boundary, which has to
be approved by the Council of State, the proposed development
does not qualify for the legal status of new town.

The syndicat communautaire was offered to the communes as
an alternative to the existing possibilities of merging their
interests in a single new commune (ensemble urbain) or of
working within the structure of a communauté urbaine where
individual communes are represented on the wider council of
the communauté.   Only in the case of Le Vaudreuil has the former solution been adopted, whilst Lille-Est is the only new
town the building of which has been directed by a communauté
urbaine.   The other seven have made use of the procedure of
the syndicat communautaire.

The SCA has been favoured by local mayors because it enables them to retain a measure of control over the government of
their respective communes whilst collaborating with their

neighbours in managing the building of the new town.   Under
the SCA arrangement a zone d'agglomération nouvelle (ZAN) is
created, mostly outside the limits of already-urbanized areas,
and within this zone the more important powers of local government (especially finance) are conferred on the new (SCA) authority although residents continue to elect councillors to serve
the existing communes.   Representatives of these councillors
serve on the new authority, the 'comité syndical', a kind of
secondary municipal council, and the new town has a president
instead of a mayor.   Under the Loi Boscher it is envisaged
that this form of local control over how the new town is directed will cease after 25 years when the ZAN assumes the status
of a commune in its own right.   Meanwhile the sensibilities
of local officials are respected.   The degree to which this
is the case is best represented in the example of Melun-Sénart
where, because the new development is proceeding in areas that
are discontinuous, it has been found necessary to set up three
syndicats communautaires.
     The local government powers of the SCA are not sufficient
to undertake the task of equipping the new town with roads,
sewers, water supply and other elements of basic infrastructure.
This work is undertaken by a technical body, the établissement
public d'aménagement (EPA).   An EPA is set up after the
mission d'études has completed the initial planning and a
master plan for the new town has been approved, its staff
usually being the same as that of the mission to ensure continuity between plan-making and implementation.   Parallels
have been drawn between the role of the EPA and that of the
development corporation of a British new town.   As prime
developers they certainly have much in common, but the EPA's
powers are less complete than those of its British counterpart,
its proposals having to be approved by the SCA, for example,
whilst the development corporation is less fettered by existing
local government structures.
     There is, however, greater flexibility in the financing
of new towns in France than is the case in Britain.   Funds
are conferred on the EPA by central government and finance is
also raised in the form of loans from the Caisse des Dépôts
et Consignations and other bodies as well as from the later
sale of land.   With these funds the EPA seeks to acquire land
at low cost, taking advantage of the ZAD procedure, and then
carries out the work of equipping the town with its basic services.   This done, the EPA will either lease or sell land for
development.   Leasing is most common in the town centre but
land is commonly sold for housing or shops in other parts of
the town.   This will follow its division into zones d'aménagement concerté where contractual agreements can be entered into
between the public authority (in this case the EPA) and representatives of the private sector (Chapter 5).   ZACs are
also declared in the town centre, although here the develop-

ment is likely to be managed by the EPA.

EPAs were set up in all nine of the new towns between 1969 and 1973. Each one is presided over by an administrative council which is made up of representatives of the government, local authorities and other 'qualified persons'. Relations with the SCA vary, despite the fact that the two bodies may have the same presiding head. Where they have been difficult it has resulted from the ingrained suspicion which local councillors tend to have for experts from outside. Diffi- culties have also arisen as a result of the varied political complexion of the communes making up the new town. Evry got off to an early start by leaving out of the designated area those communes which had communist administrations (Rubenstein, 1978). The fact that the boundaries of the SCA and EPA are not coincident also makes it more difficult to achieve a new town identity; indeed there is sometimes confusion as to the precise extent of the new town. Uncertainties of this kind are not helped by changes that have been made to the projected population totals of the new towns. Such problems as these are best appreciated by considering each town individually.

Saint-Quentin-en-Yvelines

Of the five towns retained in the 1969 revision of the Schéma for the Paris Region, Saint-Quentin-en-Yvelines has come closest to achieving the housing targets set for it, though even this falls far short of the ambitious hopes held out for the new towns in 1965. It was one of the first to acquire an EPA (October, 1969) and between that time and 1977 its 'new town' population grew from 4,000 to 48,000 (Pitié, 1978). By the latter date the population of the eleven communes which had contributed some or all of their territory to the SCA had reached 120,000 (23,650 in 1962). By 1982 the total had risen to 150,000.

The plan for Saint-Quentin-en-Yvelines envisages a series of sub-towns (bourgades) of between 25,000 and 40,000 inhabi- tants grouped around an urban core with a population of about 100,000. Work on this core zone is well advanced but progress on the bourgades has been slower with only that of Elancourt- Maurepas nearing completion. In contrast with several of the other new towns in the Paris Region, the planners of Saint- Quentin-en-Yvelines have not laid stress on the early completion of a shopping centre that might serve to give it an 'image'. Instead the emphasis has been on housing in a well-landscaped setting that includes a 120 ha. lake. Open spaces have helped to create the town's reputation as 'la ville à la campagne', and a third of the housing takes the form of individual dwellings Thirty kilometres to the south-west of Paris, and 15 km from Versailles, Saint-Quentin-en-Yvelines is in the fashionable portion of the suburbs and this has contributed to the town's success, including its attraction to employers, both factory-

and office-based.

By 1980, work on the rest of the town was sufficiently advanced to justify a new shop and office centre, and the design of this was entrusted to the Spanish architect, Manolo Nuñez. His plan draws its inspiration from the medieval town with gateways leading to a central place surrounded by galleried shops.

Cergy-Pontoise

The new town of Cergy-Pontoise is being built around a double meander of the river Oise, 25 km north-west of Paris. The EPA dates from April, 1969 and Cergy-Pontoise was the first of the Parisian new towns to experience actual construction. By 1982 the population of the 15 communes which make up the new town had reached 120,000, a third of these being newcomers.

The master plan envisaged five new quarters, work being most advanced on that of Cergy, better known as the 'Quartier de la Préfecture' since it has at its core the administrative headquarters of the département of Val-d'Oise. The distinctive inverted pyramid building of the new préfecture was opened in 1970. Since that time a 'regional' shopping centre, Les Trois Fontaines, has been completed which boasts the only department stores in the north-western suburbs, and the year this opened (1973) also saw the move to Cergy from Paris of the Ecole Supérieure des Services Economiques et Commerciales (ESSEC).

Work is also in progress on two other quarters, those of Eragny to the south and of Menucourt to the west. Others planned are close to the Forêt de l'Hautil to the south-west where the emphasis will be on individual housing but where there has been strong opposition from local residents, and at Puiseux to the north where it is hoped to build a still-larger commercial centre. Open spaces will be reserved between these developments, particularly in the actual valley of the Oise where the meander downstream from Cergy is given over to lakes and recreational uses.

The reputation of Cergy-Pontoise suffered in the early years from its relatively poor communications and from its high proportion of low income/high rise dwellings. The plan for an aérotrain link to La Défense was abandoned in 1974 and it is likely to be 1986 before the new rail link with the RER system is finally completed. The A.15 motorway has likewise taken many years to construct. Like the other new towns, Cergy-Pontoise has a large number of young families and the demand has been strong for various forms of 'aided' housing. Attempts have been made, however, to avoid building too high a proportion of the cheap (très aidé) housing which characterized the grands ensembles and there is an increasing proportion of HLM housing being built for owner-occupiers (aidé). Greater numbers of individual houses are also being built, particularly in the 'old' town nuclei such as Pontoise.

Marne-la-Vallée

Strung out along the south bank of the Marne, straddling
two départements and 21 communes, Marne-la-Vallée exemplifies
well the problems of building a new town within the framework
of existing administrative structures.   It is also in the
eastern suburbs, and although Noisy-le-Grand at the western
end of the new town is only ten km from the Boulevard
Périphérique, served by the Autoroute de l'Est and, since 1977,
by the Réseau Express Régional, the new town has been less successful than either Saint-Quentin-en- Yvelines or Cergy-Pontoise
in achieving the planners' aim of 'un emploi, un logement'.
Ever since the government made known its intention of moving
some of the departments of the Ministry of Finance from central
Paris to the new town, it has been advertised as the 'cité
financière bis' of the Paris Region, but there is little to
show for this policy beyond the arrival of the Ecole Nationale
du Trésor in 1978.

The new town is divided into three sections each of which
demands a separate syndicat communautaire d'aménagement.   The
population of the designated area was approaching 120,000 in
1982 but the 'new town' part of this total was mainly confined
to the western-most section where a new commercial centre,
the Piazza, was opened at Noisy-le-Grand in 1978.   Housing
here is mostly in the form of apartments but plans for the
eastern-most section of the new town, close to Lagny, reflect
the present-day desire for low-density individual houses.

Evry

The site of Evry is on a low plateau some 30 km to the
south-east of Paris.   The 14 communes that were the subject
of the feasibility study for the new town were reduced to five
when nine of them refused to join on the grounds that the SCA
was undemocratic and that the EPA was dominated by outside
technocrats.   Evry survived this trauma, setting up its EPA
in 1970 and has marketed itself well to become the best known
of the Parisian new towns.   It houses the préfecture of the
département of Essonne and achieved considerable publicity
when a government cabinet meeting was held here in February,
1975, billing itself as the capital of France for one day.

The plan of the new town takes the form of a figure '8',
linear neighbourhoods converging on the town centre.   The
latter includes, in addition to the buildings of the préfecture,
a major shopping centre and the Agora, a town square flanked
by buildings devoted to cultural and leisure use.   Hotel and
conference facilities have helped to promote the image of the
new town.   The population of the five communes, only 9,400
in 1968, had risen to 37,000 by 1982.   Early housing was
innovative and included the pyramid-shaped apartment blocks of

Evry I, designed so as to provide terraces for the individual housing units.   As in the other new towns, more recent developments have tended to include a higher proportion of individual houses.   The various neighbourhoods are separated by areas of parkland.

Evry is well served by modern communications, being close to the A.6 (the 'Autoroute du Soleil'), and the older route nationale 7.   A new rail link with the Gare de Lyon has been built which, since 1977, feeds into the RER system.   The town is also conveniently near to Orly airport.   Good transport facilities have proved attractive to employers and some 15,000 jobs were created in the new town during the 1970s. The opening of the Institut Universitaire de Technologie in 1972 marked the first phase in the creation of what is intended to be the university of Evry-Melun-Sénart.

Melun-Sénart

Melun-Sénart is, in contrast with Evry, the least innovative of the new towns of the Paris Region.   It was the last to be started, its EPA dating from October, 1973, and it has built the fewest houses.   Like Marne-la-Vallée, Melun-Sénart has the problem of dealing with two départements and a large number of communes (18), and three separate SCAs are involved in the development of the new town.   Thirty-five kilometres from the Boulevard Périphérique, it is further from Paris than any of the other new towns, but this does afford the advantage of cheaper building land.   That, in turn, enables the town to offer a higher proportion of individual family houses than has been typical elsewhere.   So far much of the development has taken the form of extension to existing settlements, the 18 communes having a population of 120,000 in 1982, only a small proportion of which was attributable to the 'new town' component.

Le Vaudreuil

The new town of Le Vaudreuil, 25 km south-east of Rouen at the confluence of the river Eure with the Seine, is a product of plans for the lower Seine valley drawn up in the mid to late 1960s.   A mission d'études had been appointed in 1965 to carry out a study of the lower valley and the site of Le Vaudreuil is identified in its schéma d'aménagement published in 1967.   The Seine valley was recognized at this time as one of the zones d'appui (support zones) which were most likely to receive overspill from the Paris Region and the purpose of the new town was to concentrate growth, avoiding the sprawling, linear kind of development that might otherwise occur between Paris and Rouen.

An EPA was set up in 1972 and the first newcomers were housed in 1975 but growth has been slow, averaging no more than 400-500 new dwellings a year, so that the population of

the new town totalled only 5,500 in 1981. In the heady,
expansionist years of the 1960s the new town was expected to
have a population of 140,000 by the end of the century. Present plans are for a modest 20,000 by the year 1990. About
3,000 jobs have been created in the new town, more than there
are working residents at present to fill them.

The 3,400 hectares which make up the site of the new town
have been carved out of eight rural communes and Le Vaudreuil
is unique in its achievement of sufficient harmony to create
an ensemble urbain. This is now a commune in its own right
and has a mayor.

Lille-Est

Lille-Est differs from the other new towns in its proximity
to existing urban centres. It is only seven km from the middle of Lille and a similar distance from Roubaix. It is
also heavily dependent on the university and science park
which form the nucleus of the new town.

The decision to create an educational and research campus
to the east of Lille was taken in the late 1950s and the first
buildings were opened on the site in 1963. By 1967, when it
was decided to build a new town to serve the complex, plans
were already advanced for the establishment here of new
faculty buildings for the university, first for mathematics
and science, later for arts subjects. An EPA was set up in
April, 1969 and work on the new town began in 1971.

The purpose behind the new town of Lille-Est has been to
create an attractive living environment and service infrastructure for the various laboratories, research and teaching
establishments that now cluster on the site. These include,
in addition to the university, the Centre de Recherches et
d'Etudes Supérieurs du Textile (CRETS) which serves the French
textile industry, the Centre d'Etudes et de Recherches de
Technologie des Industries Alimentaires (CERCIA) which fulfils
a similar role in relation to the food and drink industries,
the Institut National de la Recherche de Chimie Appliquée
(IRCHA) and the Institut de la Santé et de la Recherche Médicale
(INSERM). In constructing the new town, considerable attention has been paid to landscaping, including the creation of
lakes and tracts of woodland as well as the preservation of
old farm buildings in the Flemish style. Not surprisingly,
the residential development includes areas of expensive single-
family housing as well as the low apartment building more
typical of the new towns in general. Another feature of the
new town is the regional-scale sports complex which includes
a stadium, swimming pools and sailing lake of 'Olympic' proportions.

Some 26,000 people lived on the site of the new town at
the time of designation. That total has since risen to
60,000 and although the original intention was to create a

town of 100,000, possibly even 150,000 inhabitants, there seems
to be a possibility now that planned growth will cease when
the population reaches 65,000 in 1984-85. The town's
greatest problem lies in achieving an identity of its own rather than being regarded simply as a scientific appendage of
greater Lille.   With this in mind, the inhabitants prefer to
use the town's alternative name - Villeneuve-d'Ascq - which is
derived from that of one of the three communes which combined
(under the legislation of July, 1971) to form a new commune to
serve as site for the new town.   This commune is a part,
however, of the communauté urbaine of Lille which retains the
primary responsibility for the planning of the new town.   The
close link with Lille is symbolized in the opening of the light
métro, expected in 1983 (Chapter 3).   The acronym 'VAL', some
insist, stands for Villeneuve-d'Ascq-Lille.

L'Isle-d'Abeau

Conceived in the late 1960s, L'Isle-d'Abeau was to have
been one of two new towns serving the Lyon area.   It is a
typical product of that period when the expectation was of
continued demographic and economic growth and when free rein
was given to theories of spatial planning.

Thirty-three kilometres from the centre of Lyon on the
A.43 motorway which forms part of the route axis to Grenoble,
this 'petite soeur de Lyon' is in a different département from
the capital of Rhône-Alpes and has always been viewed with
suspicion by its older sister (Chapter 3).   The attitude to
Le Vaudreuil in Rouen is, in fact, very similar.   An EPA was
constituted in 1972 to build L'Isle-d'Abeau and the intention
then was that the new town would achieve a population total of
120,000 by 1985 and possibly 200,000 by the turn of the century.   Its growth would be stimulated by the needs of the new
international airport of Satolas, then under construction a
few kilometres away.

Progress has been much slower than planned and the population of the eight communes which make up the SCA was little
more than 20,000 in 1982.   In part this has been due to the
slow pace of job creation at the new airport, and an eventual
population of 60,000-80,000 now seems more probable than the
totals projected in the early 1970s.   The plan for L'Isle-
d'Abeau is a loose-knit one, based on a series of linked
villages (bourgades) each planned to house some 25,000-30,000
inhabitants and to be divided into four or five smaller neighbourhoods.   These are to be joined by parkways to the town
centre, much of the construction being low density with a high
proportion of individual houses and a country setting ('une
ville verte').   Only one district, that of Villefontaine in
the south-west of the new town is as yet advanced and it
remains to be seen whether the plan for a 'fédération de
bourgades' can survive the secession of several communes from

the original designated area and the reduction in the overall
ambitious plans for the new town which have taken place in
recent years.

Les Rives de l'Etang de Berre

The new town being built on the shores of the Etang de
Berre, west of Marseille, is intended to complement the port
and industrial complex of Fos for whose workers it provides
housing and services.   The purpose is clear, but the new town
itself is a curious hybrid which illustrates vividly the problems that arise when inter-communal collaboration is sought.

The Etang de Berre is bordered by a dozen or so small
townships of varied political complexion.   In broad terms
those to the east and the north-west tend to be more centrist
in their outlook than the ones in the south-west, near the entrance to the Etang, where support for the Communist Party is
traditionally strong.   Hopes that all the communes might be
united in the new town project foundered on this question of
politics.   Three communes - Miramas, Istres and Fos - which
in the early 1970s had councils sympathetic to the government,
formed a syndicat communautaire d'aménagement in 1972 and these
three, plus the commune of Vitrolles in the east, are the concern of the établissement public d'aménagement set up in 1973.
But three of the more left-wing communes - Martigues, Port-de-
Bouc and Saint-Mitre - are also interested in the new town
development and have formed themselves into a syndicat intercommunal à vocation multiple (SIVOM).   Central government
recognizes them as a part of the new town and various other
organizations have been established in order to achieve coordination in planning between the different groupings of communes.   These include a syndicat mixte de coordination and a
mission ministérielle pour l'aménagement de la région de Fos-
Etang de Berre; whilst the département, OREAM and Marseille
port authority are all involved in other ways.

Some 9,000 new jobs had been created in the docks and
factories of Fos by 1979, well below the projected total of
30,000.   The slow growth of Fos, which owes much to the
recession, has reduced the pressure on the new town where most
of the new housing has been built in Miramas and Istres.   By
1982 the population of the four communes which make up the EPA
had risen to 80,000.

Prospects for the New Towns

The new towns are seen by their critics as symbols of the
1960s.   To them, they are typical products of a period of
demographic and economic euphoria that placed its faith in
'technocratic' solutions to human problems.

The scale on which individual new towns were conceived
lends support to this view.   But since the first plans were
published their target populations have, in most cases, been

severely reduced, making them less vulnerable to accusations of 'gigantisme'. Furthermore, the rate of housebuilding has been much less than first envisaged, the Parisian new towns achieving only 44% of their planned totals during the period of the Sixth Plan (1971-75). The four provincial towns did rather better, reaching 70% of their goal overall. Tuppen (1979) sees the failure of the Parisian new towns to reach their target populations as a reflection of the basic inconsistency of planning for large-scale growth in Paris whilst at the same time carrying out policies aimed at decentralization. He cites the necessity of paying a development tax (redevance) in the new towns as an example of this inconsistency.

Another possible reason for disaffection lies in the slowness of the new towns to adapt their plans to changing tastes in housing, their budgets not allowing them the luxury of building large areas of low density pavillons. This may help to explain why the new towns contributed only 11.5% of the housing built in the Paris Region between 1971 and 1975 compared with the 24% envisaged in the Sixth Plan (Carmona, 1977). A consequence of this slow pace of growth is the underuse of services already provided in the new towns.

On the credit side the new towns have provided a living environment vastly superior to that found in the grands ensembles. Their architecture is infinitely more varied and great care has been taken over landscaping and the provision of amenities of all kinds including facilities for sport and other leisure activities. The population living in the nine new towns had risen from 252,000 in 1968 to 540,000 by 1980 (or 730,000 if a broader definition is taken embracing all the communes involved). The number of new dwellings built during this period totalled 137,000.

Left-wing opponents of the new towns have long objected to what they see as the 'undemocratic' arrangements made for directing their growth under the Loi Boscher of 1970. To them the syndicat communautaire d'aménagement represents state-domination over local democracy, and there have been calls to end the system by which the new town is administered and financed by a separate organization from the communes of which it may be a part. Legislation put before the National Assembly in October, 1982 seeks to satisfy these critics by 'reforming' the Loi Boscher, and after the municipal elections of 1983 the local authorities concerned are to be offered a number of alternative solutions. These include the possibility of creating an entirely new commune, a syndicat intercommunal, or a communauté d'agglomérations nouvelles with a directly-elected council. Communes would no longer be divided between village and new town sections, those who wished to withdraw from the arrangement being permitted to do so. As in all urban matters in France, the political theme is never far from the surface.

Chapter 10.

POSTSCRIPT

France has been transformed within the space of little
more than thirty years from a semi-rural to an urban society.
The effects of this transformation are everywhere apparent as
historic city centres have been engulfed in a suburban tide of
housing, factories, superstores and expressways. Whether
their home is a bungalow or an apartment in a high-rise block
of flats, the physical conditions under which most French families now live is infinitely superior to those endured during
the long years of inactivity and neglect before the Second
World War. But, as Paul Claval (1981) observes, 'le citadin
est plus complexe que l'homme des villes'. The social adjustment to urban living can be a slow process for migrants from
the country or abroad, whilst the move from city to suburb can
be equally traumatic for those nurtured in the supportive
social webs of the old quartier. 'Les gens sont devenus des
nomades' (President Mitterrand in a television broadcast,
December, 1981).
    This sense of rootlessness, of disaffection with the
dreary suburbs, of isolation from city centres, is what, according to some observers, lies behind the shift to the political
left in France (Chapter 1). Frédéric Gaussen (1981) put it
plainly when he wrote, 'la gauche a profité du mécontentement
provoqué par l'urbanisation accélérée ... les germes de révolte
prolifèrent dans le terrain malsain des grands ensembles'. It
was 'le vote du béton'(concrete).
    Clearly cities must differ in this respect, however. In
a study of three cities of the Loire valley, Perrineau (1981)
contrasts the situation in Tours where the integration of the
old and new parts of the city has been relatively successful,
with that in Orléans and Blois where large, ill-equipped ZUPs
languish uneasily on the urban fringe. Here the inhabitants
exhibit 'un sentiment de frustration et d'exclusion' which has
manifested itself in a marked swing to the political left. In
the better-planned city of Tours, M.Royer was re-elected mayor
despite his reputation as an ultra-conservative.

The rapid growth of French cities over the last thirty years has created problems as well as solving them. Houses and services have been provided en masse but at the expense of character and of those very qualities of urbanity which the city is meant to symbolize. The greater interest now being shown, both in social issues and in matters of conservation, is a hopeful sign, but much remains to be done if the French are to respond to President Mitterrand's call to lay the foundations of a 'nouvelle civilisation de la ville'.

BIBLIOGRAPHY

Alonso, W. (1971) 'The Economics of Urban Size'. Papers of
  the Regional Science Association, 21, 67-83
Ardagh, J. (1980) France in the 1980s. Harmondsworth: Penguin
  Books
Aydalot, P. (1976) Dynamique Spatiale et Développement Inégal.
  Economica, Paris
Barbier, B. (1978) 'La Consommation d'Espace Liée à la
  Croissance Urbaine en France'. In: Kobori, I. (Ed.)
  Colloque sur la Croissance Urbaine en France et au Japon.
  Tokyo: Japan Society for the Promotion of Science, pp.49-55
Barrère, P. and Cassou-Mounat, M. (1980) Les Villes
  Francaises. Paris: Massou
Bastié, J. (1964a) La Croissance de la Banlieue Parisienne.
  Paris: Presses Universitaires de France
———— (1964b) Paris et l'An 2000. Paris: SEDIMO
———— (1975) Vingt Ans de Transformation de Paris: 1954-
  1974. Paris: Association Universitaire de Recherches
  Géographiques et Cartographiques
Beaujeu-Garnier, J. (1975) La Population Française. Paris:
  A.Colin
Beaujeu-Garnier, J. and Bouveret-Gauer, M. (1979) 'Retail
  Planning in France'. In: Davies, R.L. Retail Planning in
  the European Community. Farnborough: Saxon House, pp.99-112
Bédarida, F. (1980) 'Towns in England and Wales'. In: Johnson,
  D., Crouzet, F. and Bédarida, F. (Eds) Britain and France:
  Ten Centuries. Folkestone: Dawson, pp.225-233
Beresford, M.W. (1967) New Towns of the Middle Ages: Town
  Plantation in England, Wales and Gascony. London: Lutterworth
Bienfait, M. (1973) 'Immigration et Emigration dans l'Agglo-
  mération d'Albi de 1962 à 1968'. Rev. Géog. des Pyrénées et
  du Sud-Ouest, 44, 485-492
Bouchet, J. and Muron, J.-L. (1978) 'Enseignements du Recense-
  ment et Conditions Nouvelles pour l'Aménagement du Territoire'.
  In: Les Disparités Démographiques Régionales. Paris: Editions
  du Centre National de la Recherche Scientifique.

Boudeville, J.-R. (1958) L'Economie Régionale: Espace
Opérationnel. Cahiers de l'ISEA, Série L, No.3
Brayne, M.L. (1972) 'Rungis: the New Paris Market'.
Geography, 57, 47-51
Brissy, Y. (1974) Les Villes Nouvelles: le Rôle de l'Etat et
des Collectivités Locales. Paris: Berger-Levrault
Cabanne, C. (1979) 'L'Evolution de la Construction Navale à
l'Estuaire de la Loire'. Norois, 104, 499-506
Carmona, M. (1975) 'Les Plans d'Aménagement de la Région
Parisienne'. Acta Geographica, 23, 15-46
——————— (1977) 'Les Villes Nouvelles de la Région Parisienne'.
Acta Geographica, 29, 25-52
Carrière, F. and Pinchemel, P. (1963) Le Fait Urbain en France.
Paris: A.Colin
Cassou-Mounat, M. (1978) 'L'Evolution Récente des Structures
Commerciales dans l'Agglomération de Bordeaux'. Rev. Géog.
des Pyrénées et du Sud-Ouest, 49, 75-98
Castells, M. (1978) City, Class and Power. English trans-
lation by Elizabeth Lebas. Basingstoke: Macmillan
Chabot, G. (1960) Présentation d'Une Carte a 1:625,000 des
Zones d'Influence des Grandes Villes Françaises. 19th
Congrès International de Géographie
Chaline, C. (1980) La Dynamique Urbaine. Paris: Presses
Universitaires de France
Charles, G. (1979) 'Deux "Villes Moyennes" de Franche-Comté:
Dole et Vesoul'. Rev. Géog. de l'Est, 19, 365-370
Charrier, J.-B. (1977) 'Chronique Bourguignonne: Décize et la
Région Décizoise'. Rev. Géog. de l'Est, 17, 215-229
Choay, F. (1965) L'Urbanisme: Utopies et Réalités. Paris:
Editions du Seuil
Claval, P. (1981) La Logique des Villes: Essai d'Urbanologie.
Paris: Litec
Clerc, P. (1967) Grands Ensembles, Banlieues Nouvelles.
Paris: Centre de Recherche d'Urbanisme
Clout, H.D. (1977a) 'Early Urban Development'. In: Clout,
Hugh D. (Ed.) Themes in the Historical Geography of France.
London: Academic Press, pp.73-106
——————— (1977b) 'Urban Growth, 1500-1900. In: Ibid.,
pp.483-540
Cohen, S.S. (1978) 'Paris'. In: Cités Géantes. Paris: Fayard
Comby, J. (1973) 'Un Nouvel Aspect de la Politique de la DATAR:
les Villes Moyennes, Pôles de Développement et d'Aménagement?'.
Norois, 20, 647-660
——————— (1974) 'L'Opération "Angoulême, Ville Moyenne Pilote":
la Fin d'une Illusion?'. Norois, 21, 497-504
——————— (1975) 'Trente Ans d'Urbanisme dans Cinq Villes
Moyennes'. CRU Annales, pp.31-66
Coppolani, J. (1959) Le Réseau Urbain en France: Sa Structure
et Son Aménagement. Paris: Les Editions
Ouvrières

Dalmasso, E. (n.d.) 'France', The National Urban Systems. International Geographical Union, Commission on National Settlement Systems, unpublished typescript

Darin-Drabkin, H. (1977) Land Policy and Urban Growth. Oxford: Pergamon

Dawson, J.A. (1976) 'Hypermarkets in France'. Geography, 61, 259-262

————— (1981) 'Shopping Centres in France'. Geography, 66, 143-146

Documentation Française, La (1976) Les Petites Villes en France. Recherches de Prospective, Paris

Duclaud-Williams, R.H. (1978) The Politics of Housing in Britain and France. London: Heinemann

Dumas, J. (1978) 'Transformations Industrielles et Intégration de l'Espace dans la Communauté Urbaine de Bordeaux'. Rev. Géog. des Pyrénées et du Sud-Ouest, 49, 51-74

Dyer, C. (1978) Population and Society in Twentieth Century France. London: Hodder and Stoughton

Estienne, P. (1979) La France: I. Généralités, Région du Nord. Paris: Masson

Evenson, N. (1979) Paris: A Century of Change, 1878-1978. New Haven

Fielding, A.J. (1966) 'Internal Migration and Regional Economic Growth: A Case Study of France'. Urban Studies, 3, 200-214

Florence, P.S. (1948) Investment, Location and Size of Plant. Cambridge: University Press

Gaussen, F. (1981) 'Le Vote du Béton'. Le Monde Dimanche, 1 November, p.3

Geoffroy, B. (1980) 'La Naissance d'un Espace Péri-urbain en Province: Le Cas de La Roche-sur-Yon (Vendée). Norois, 108, 609-614

George, P. (1952) La Ville. Paris: Presses Universitaires de France

Gottmann, J. (1976) 'Paris Transformed'. Geog. Journal, 142, 132-135

Gravier, J.-F. (1947) Paris et le Désert Français. Paris: Le Portulan

————— (1958) 'Réalité de la Région'. Urbanisme, No.58

Hall, P. (1966) 'Paris'. In: Hall, Peter (Ed.) The World Cities. London: Weidenfeld and Nicolson, pp.59-94

Hans, M.E. (1974) 'Les Conditions du Logement au Centre des Agglomérations'. Economie et Statistique, 55, table 5

Hautreux, J. (1972) Les Principales Villes Attractives et leur Ressort d'Influence. Paris: Centre d'Etudes d'Aménagement et d'Urbanisme

Hautreux, J. and Rochefort, M. (1963) Le Niveau Supérieur de l'Armature Urbaine Française. Paris: Ministère de la Construction

————— (1964) La Fonction Régionale dans l'Armature Urbaine

<u>Française</u>.   Paris: Ministère de la Construction
─────── (1965) 'Physionomie Générale de l'Armature Urbaine
Française'. <u>Annales de Géographie</u>, 74, 660-677
Idrac, M. (1973) 'Les Problèmes de Développement de Montauban'.
<u>Rev. Géog. des Pyrénées et du Sud-Ouest</u>, 44, 397-414
─────── (1979) 'Commerce et Aménagement Urbain: Le Cas de
l'Agglomération de Toulouse à la Fin des Années 1970'.
<u>Rev. Géog. des Pyrénées et du Sud-Ouest</u>, 50, 7-26
Jeanneau, J. (1974) 'Le Processus de Déconcentration Urbaine:
L'Exemple d'Angers'. <u>Norois</u>, 81-84,427-441
─────── (1978) 'Cholet et Saumur: Deux Politiques d'Aménage-
ment du Centre en Ville Moyenne'. <u>Norois</u>, 97-100, 87-101
Jones, P.N. (1978) 'Urban Population Changes in France,
1962-75'. <u>Erdkunde</u>, 32, 198-212
Kain, R.J. (1982) 'Europe's Model and Exemplar Still? The
French Approach to Urban Conservation, 1962-1982'. <u>Town
Planning Review</u>, 53, 403-422
Katan, Y. (1981) <u>Paris et la Région Ile-de-France</u>. Paris:
Hatier
Kayser, B. (1973) 'Croissance et Avenir des Villes Moyennes
Françaises'. <u>Rev. Géog. des Pyrénées et du Sud-Ouest</u>, 44,
345-364
Kinsey, J. (1979) 'The Algerian Movement to Greater Marseille'.
<u>Geography</u>, 64, 338-341
Labasse, J. (1974) <u>L'Espace Financier</u>. Paris: A.Colin
Laferrère, M. (1970) 'Les Fonctions Tertiaires d'une Grande
Ville de Province'. In: <u>Grandes Villes et Petites Villes</u>.
Colloques Nationaux du Centre National de la Recherche
Scientifique, pp.199-214
Lajugie, J. (1974) <u>Les Villes Moyennes</u>. Paris: Editions Cujas
Lajugie, J., Delfaud, P. and Lacour, C. (1979) <u>Espace Régional
et Aménagement du Territoire</u>. Paris: Dalloz
Lévy, J.-P. (1973) 'La Croissance d'une Ville Tertiaire, Auch'.
<u>Rev. Géog. des Pyrénées et du Sud-Ouest</u>, 44, 429-443
─────── (1977) 'Le Mirail en 1977'. <u>Rev. Géog. des Pyrénées
et du Sud-Ouest</u>, 48, 103-114
Lévy, J.-P. and Poinard, M. (1973) 'Dynamique de l'Emploi
Tertiaire dans les Villes Moyennes de Midi-Pyrénées'. <u>Rev.
Géog. des Pyrénées et du Sud-Ouest</u>, 44, 365-382
Lugan, J.-C. and Poinard, M. (1973) 'Les Orientations d'une
Petite Capitale Régionale: Rodez'. <u>Rev. Géog. des Pyrénées
et du Sud-Ouest</u>, 44, 415-428
Merlin, P. (1969) <u>Les Villes Nouvelles</u>. Paris: Presses
Universitaires de France
─────── (1971) <u>Vivre à Paris 1980</u>. Paris: Hachette
Metton, A. (1980) 'Les Mutations de l'Equipement Commercial:
Un Aspect de l'Evolution Urbaine'. <u>Norois</u>, 108, 601-608
─────── (1982) 'L'Expansion du Commerce Périphérique en
France'. <u>Annales de Géographie</u>, 506, 463-479
Metton, A. and Meynier, A. (1981) 'A Propos des Rues Piéton-

nières'. Norois, 109, 65-72

Michel, M. (1977) 'Ville Moyenne, Ville-Moyen'. Annales de Géographie, 86, 641-685

Noin, D. (1976) L'Espace Français. Paris: A.Colin

Ogden, P.E. and Winchester, S.W.C. (1975) 'The Residential Segregation of Provincial Migrants in Paris in 1911'. Transacs. Inst. Br. Geogrs., 65, 29-49

Osborn, M. (1967) 'The Frenchman's Home'. Town and Country Planning, 35, 366-368

Pagès, M. (1980) La Maitrise de la Croissance Urbaine. Paris: Presses Universitaires de France

Palu, P. (1975) 'Les Structures Commerciales de Tarbes'. Rev. Géog. des Pyrénées et du Sud-Ouest, 46, 297-311

————— (1982) 'Les Politiques Commerciales en Centre-ville'. Annales de Géographie, 506, 435-441

Parker, A.J. (1975) 'Hypermarkets, the Changing Pattern of Retailing'. Geography, 60, 120-124

Perrineau, P. (1981) Espace Politique: Les Conséquences Politiques du Changement Urbain dans les Agglomérations de la Loire-Atlantique (1958-1978). Thèse de Doctorat de l'Etat, Institut d'Etudes Politiques de Paris

Perroux, F. (1950) 'Les Espaces Economiques'. Economie Appliquée, Archives de l'ISEA, No.1, pp.225-244

Piatier, A. (1956) 'L'Attraction Commerciale des Villes, une Nouvelle Méthode de Mesure'. Revue Juridique et Economique du Sud-Ouest, 4, 575-594

Pinchemel, P. (1969) France: A Geographical Survey. London: Bell

————— (1979) La Région Parisienne. Paris: Presses Universitaires de France

Pitié, J. (1978) 'Les Villes Nouvelles Françaises en Question'. Norois, 99, 453-460

————— (1979) 'Poitiers à travers Quelques Travaux Récents'. Norois, 104, 541-547

Poittier, M. (1974) 'Présentation d'une Série d'Enquêtes sur les Magasins à Grande Surface de Lorraine'. Bull. de l'Assoc. de Géographes Français, 413-414, 15-28

Prager, J.-C. (1973) 'L'Industrie dans les Villes Moyennes de la Région Midi-Pyrénées'. Rev. Géog. des Pyrénées et du Sud-Ouest, 44, 383-396

Pred, A.R. (1973) Systems of Cities and Information Flows. Lund Studies in Geography, Series B, No.38

Prost, M.-A. (1965) La Hiérarchie des Villes en Fonction de leurs Activités de Commerce et de Service. Paris: Gauthier-Villars

Prud'homme, R. (1973) 'Costs and Financing of Urban Development in France'. Urban Studies, 10, 189-198

Pumain, D. and Saint-Julien, T. (1978) Les Dimensions du Changement Urbain: Evolution des Structures Socio-économiques du Système Urbain Français de 1954 à 1975. Paris: Edition du Centre National de la Recherche Scientifique

Rapoport, A. (1968) 'Housing and Housing Densities in France'. Town Planning Review, 39, 341-354

Raymond, M.G., Haumont, N, Raymond, H. and Haumont, A. (1966) Les Pavillonnaires, La Politique Pavillonnaire, L'Habitat Pavillonnaire. Paris: Centre de Recherche d'Urbanisme

Rochefort, M. (1957) 'Méthodes d'Etudes des Réseaux Urbains'. Annales de Géographie, 66, 125-143

Rubenstein, J.M. (1978) The French New Towns. Baltimore: The Johns Hopkins University Press

Scargill, D.I. (1974) The Dordogne Region of France. Newton Abbot: David and Charles

Schiray, M. and Elie, P. (1970) Les Migrations Entre Régions et au Niveau Catégories de Communes de 1954 a 1962. Collections de l'INSEE, série D4

Simmie, J.M. (1981) Power, Property and Corporatism: The Political Sociology of Planning. Basingstoke: Macmillan

Simonetti, J.-O. (1978) 'Reflexions sur l'Industrialisation de la Construction et la Production du Bâti'. Norois, 95-96. 341-353, 561-572

Siran, J.-L. (1978) Nouveaux Villages, Nouvelles Banlieues. Paris: Sorbonne

Smith, B.A. (1973) 'Retail Planning in France: The Changing Pattern of French Retailing'. Town Planning Review, 44, 279-306

Soumagne, J. (1977) 'Le Commerce de Détail de l'Agglomération de La Rochelle'. Norois, 93-94, 53-82, 193-210

Steinberg, J. (1976) 'Caractères et Conditions d'Implantation de Deux Urbanisations Nouvelles de l'Est Parisien: le Nouveau Créteil et Noisy-le-Grand/Marne-la-Vallée'. La Vie Urbaine, 51, 23-44

Sueur, G. (1971) Lille-Roubaix-Tourcoing, Métropole en Miettes. Paris: Stock

Sutcliffe, A. (1970) The Autumn of Central Paris. London: Arnold

———— (1980) 'Towns in England and Wales'. In: Johnson, D., Crouzet, F. and Bédarida, F. (Eds), Britain and France: Ten Centuries. Folkestone: Dawson, pp.217-224

Thompson, I.B. (1981) The Paris Basin: Problem Regions of Europe. Oxford: University Press

Thouvenot, C. and Wittmann, M. (1970) 'La Métropole Lorraine'. L'Information Géographique, No.3, pp.107-117

Tulet, J.-C. (1973) 'Nature de la Croissance Economique de Cahors'. Rev. Géog. des Pyrénées et du Sud-Ouest, 44, 445-460

Tuppen, J. (1979) 'New Towns in the Paris Region: An Appraisal'. Town Planning Review, 50, 55-70

———— (1980) 'Public Transport in France: the Development and Extension of the Métro'. Geography, 65, 127-130

Vassal, S. (1977) 'Urbanisation et Vie Rurale: Le Cas de

l'Agglomération Orléanaise'. <u>Norois</u>, 95 ter (Géographie
Rurale), pp.223-238
Vaudour, N. (1974) 'L'Aire d'Attraction Commerciale d'un Hyper-
marché Aixois: Euromarché-les Milles'. <u>Méditerranée</u>, 16,
135-148
————— (1978) 'Les Grandes Surfaces Périphériques de Vente
dans les Bouches-du-Rhône'. <u>Annales de Géographie</u>, 87,
40-58
Vaughan, M., Kolinsky, M. and Sheriff, P. (1980) <u>Social
Change in France</u>. Oxford: Robertson
Vidal de la Blache, P. (1913) 'La Relativité des Divisions
Régionales'. In: Bloch, C., et al., <u>Les Divisions Régionales
de la France</u>. Paris: Alcan
Vivian, H. (1959) 'La Zone d'Influence Régionale de Grenoble'.
<u>Revue de Géographie Alpine</u>, 47, 539-583
Winchester, H.P.M. (1977) <u>Changing Patterns of French Internal
Migration, 1891-1968</u>. University of Oxford, School of
Geography Research Paper, No.17

INDEX

Milton Keynes UK
Ingram Content Group UK Ltd.
UKHW031150141024
449569UK00024B/931